CASE STUDIES IN

Clinical

Anatomy

Stuart W. McDonald

BSc, MB, ChB, PhD

Lecturer in Anatomy,
University of Glasgow
Glasgow, UK

Mosby

London   Baltimore   Bogotá   Boston   Buenos Aires   Caracas   Carlsbad, CA   Chicago   Madrid   Mexico City
Milan   Naples, FL   New York   Philadelphia   St. Louis   Sydney   Tokyo   Toronto   Wiesbaden

Copyright © 1995 Times Mirror International Publishers Limited

Published in 1995 by Mosby, an imprint of Times Mirror International Publishers Limited

Printed by J W Arrowsmith

ISBN 0 7234 2202 8

For full details of all Times Mirror International Publishers Limited titles, please write to Times Mirror International Publishers Limited, Lynton House, 7–12 Tavistock Square, London WC1H 9LB, England.

A CIP catalogue record for this book is available from the British Library.

Library of Congress Cataloging-in-Publication Data applied for.

Project Manager:	Linda Kull
Developmental Editor:	Lucy Hamilton
Layout:	Jonathan Brenchley
Cover Design:	Ian Spick
Production:	Mike Heath
Index:	Nina Boyd
Publisher:	Geoff Greenwood

PREFACE

It has been a privilege and pleasure to respond to Mosby's invitation to compile this collection of "Case Studies in Clinical Anatomy". I hope that the presentations will allow students studying Anatomy to appreciate the manner in which their knowledge will be applied in clinical practice. To encourage further reading and deeper understanding, a short bibliography accompanies each case.

I am indebted to the following colleagues for all their time and effort in so enthusiastically providing me with clinical histories; without their help this book would have been impossible.

Ms Janie S. Anderson, Rev A. Sheila Blount, Mr Colin Cameron, Dr Andrew M. Chappell, Dr Philip J. Cowburn, Ms Rosemary R. J. Fish, Ms Mary Fotheringham, Mr George A. Gillespie, Ms Gillian Kerr, Mr Brian J. McCreath, Mrs Doris P. McDonald, Mr Andrew Millar, Ms Caroline Paton, Dr Jane R. Paxton, Ms Rowan I. Shirley, Mr Fraser C. Sneddon, Mr David M. Still.

Many thanks are also owed to Mr N. Alan Green, Dr Christine H. McAlpine and Mr John G. McLennan for professional advice, to Mrs Jane Paterson for art work and to Ms Margaret Hughes for photography. I am most grateful to Ms Heather Aitken and Mr Alan J. R. Macfarlane for their comments on the text.

Stuart W. McDonald,
University of Glasgow.

DEDICATION

To my Mother and Father
and all friends from Glasgow University.

FOREWORD

I recently received letters from three young family friends who had just begun medical studies. I was impressed by the enthusiasm they all showed at the prospect of starting to learn about the form and function of the human body in the way that all medical and paramedical professionals need to know it, and I wondered what their thoughts about it would be in three months time, or three years, or ten years. Would they look back on their early studies with gratitude to their teachers and textbooks? Or would they feel disillusioned because so much of their time had been wasted on irrelevant nonsense, and that which they had been told to learn had been far too academic and unrelated to their vocational needs as custodians of the health of those under their care? In the quest for a suitable balance, Dr McDonald's book will serve an essential purpose by demonstrating the correlation between anatomical facts and their clinical applications. Of course, no one must try to run before they can walk, but an early introduction to clinical problems adds interest to acquiring the essential background knowledge that is necessary for the care of those with injury and disease. This book contains many references to further reading about the conditions discussed, and these will be most useful to students as they become more senior and begin to accumulate clinical knowledge at first hand.

For all in the health care professions there is much to learn. I commend Dr McDonald's book as one which stimulates the learning process in a fascinating fashion, and which will appeal not just to students who are completely new to the basic medical sciences but also to those who need some refreshment of those fundamental facts upon which good medical practice depends.

R.M.H. McMinn, MD, PhD, FRCS
Emeritus Professor of Anatomy
Royal College of Surgeons of England and
University of London

CONTENTS

Further reading

The page numbers cited at the top of the
Further Reading lists refer to McMinn *et al.*,
McMinn's Functional and Clinical Anatomy
(Mosby, 1995).

CHAPTER 1

HEAD AND NECK

CASE 1

Emma Blackmore, a generally healthy five-year-old girl in her first year at school, had been troubled by upper respiratory tract infections which had been treated with antibiotics. In recent weeks, her parents had noticed that she often failed to hear when spoken to. Her teacher had found this too, especially when the class was working in groups. On examining the tympanic membrane through an auriscope, Emma's doctor saw golden-brown fluid through the tympanic membrane and made a diagnosis of serous otitis media, often called 'glue ear'. The doctor allowed a medical student, on placement in the practice, to examine the ear, but told her to be very careful when inserting the auriscope.

The doctor explained to the parents and the student that glue ear was relatively frequent in primary school children and often followed a series of upper respiratory tract infections. The auditory (eustachian) tube became blocked and the epithelial lining of the middle-ear cavity developed more goblet cells than were usually present which produced the 'glue'. Emma was referred to the local paediatric unit and was subsequently admitted for aspiration of the fluid by myringotomy (a small incision in the tympanic membrane) with insertion of grommets and removal of adenoids.

QUESTIONS

1 Why did the doctor tell the student to be very careful when inserting the auriscope?
2 What type of epithelium lines the middle-ear cavity and auditory tube?
3 With which side of the middle-ear cavity does the auditory tube communicate?
4 Where does the auditory tube open into the pharynx?
5 What is the function of the auditory tube?
6 What is a grommet and what purpose does it serve?
7 What are the adenoids and why were they removed?

YOUR ANSWERS

1 Why student had to be careful inserting auriscope

2 Type of epithelium lining

3 Side of middle-ear cavity communicating with auditory tube

4 Place where auditory tube opens in nasopharynx

5 Function of auditory tube

6 Grommet

7 Adenoids and reason for removal

See McMinn pp. 169, 173, 182

Further reading
Bingham, J. G. and Hawthorne, M. R. *Synopsis of Operative ENT Surgery.* Butterworth-Heinemann, Oxford, 1992.
Bull, T. R. *A Colour Atlas of ENT Diagnosis.* 2nd edn. Wolfe Medical Publications Ltd., London, 1987.
Gray, R. F. and Hawthorne, M. *Synopsis of Otolaryngology.* 5th edn. Butterworth-Heinemann, Oxford, 1992.
Maw, T. R. Otitis media with effusion (glue ear). In: Evans, J. N. G. (ed.) *Scott-Brown's Otolaryngology: Paediatric Otolaryngology.* Butterworths, London, 1987.
de Melker, R. A. Treating persistent glue ear in children. *Br Med J* **306**, 5–6, 1993.

CASE 2

Miss Janet Allen, a 76-year-old self-employed upholsteress, was recovering from influenza when she developed pain at the right side of her forehead. Over the next few hours, the skin reddened and a vesicular rash appeared (*Fig. 1.1*). The vesicles extended over the right side of her forehead and into the hair at the front part of her head. A few vesicles also appeared on the dorsum of her nose. Her right eye and upper eyelid were involved and her vision became blurred. The left side was unaffected and the redness of the skin stopped sharply close to the midline.

When her family doctor visited, she asked the patient if she had ever had chickenpox; Miss Allen recalled being in bed with it when she was eight. A course of the antiviral drug, acyclovir, was prescribed and over the next few weeks the rash and painful inflammation resolved. Miss Allen's vision, however, never fully recovered. The blurred vision in her right eye made judging distances difficult and forced her retirement from upholstery.

QUESTIONS

1 What is the diagnosis? Why did the doctor ask about chickenpox?
2 Which nerve supplies the skin in the area affected by the rash?
3 What is the main local nerve giving cutaneous innervation to the forehead and scalp?
4 Which local nerve(s) carries sensory fibres to the dorsum of the nose?
5 What is the likely explanation for the failure of recovery of the vision in the affected eye?
6 Which local nerve(s) provides the sensory innervation of the cornea?

YOUR ANSWERS

1 Diagnosis, and why doctor asked about chickenpox

2 The nerve supplying skin in area affected by rash

3 Main local nerve innervating forehead and scalp skin

4 Local sensory nerve(s) to dorsum of nose

5 Reason for diminished vision

6 Local sensory nerve(s) to cornea

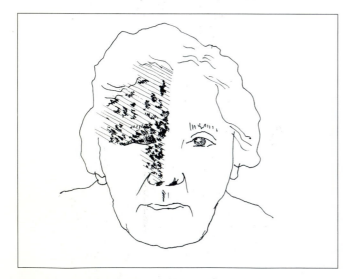

Fig. 1.1 The distribution of Miss Allen's rash.

See McMinn pp. 143, 168

Further reading
John, T. Antivirals. In: Albert, D. M. and Jakobiec, F. A. (eds.) *Principles and Practice of Ophthalmology: Basic Sciences.* W. B. Saunders Co., Philadelphia, 1994, pp. 961–971.
Pavan-Langston, D. Viral disease of the cornea and external eye. In: Albert, D. M. and Jakobiec, F. A. (eds.) *Principles and Practice of Ophthalmology: Clinical Practice.* Vol. I. W. B. Saunders Co., Philadelphia, 1994, pp. 117–161.
Weller, T. H. Varicella – herpes zoster virus. In: Evans, A. S. (ed.) *Viral Infections of Humans: Epidemiology and Control.* 3rd edn. Plenum Publishing Co., New York, 1991, pp. 659–683.

CASE 3

Mrs Elizabeth Maxwell, a 63-year-old housewife, had been troubled by vertigo (giddiness), tinnitus (ringing in the ears) and partial deafness for some weeks. Her family doctor suspected Menière's disease, a degenerative condition of the inner ear, and referred her to the ear, nose, and throat (ENT) department. After carrying out several investigations, including a computed tomography (CT) scan, the ENT specialist diagnosed schwannoma of the vestibulocochlear nerve (often called acoustic neuroma). This is a benign tumour of the Schwann cells, usually involving the vestibular component of the eighth nerve. The lesion affected the nerve of the left side and lay in the posterior cranial fossa close to the internal acoustic meatus.

Mrs Maxwell was admitted to the regional neurosurgical unit for excision of the lesion. After surgery, however, she was left with paralysis of the muscles on the left side of her face (*Fig. 1.2*). Mrs Maxwell was unable to close her left eye and required partial tarsorrhaphy (suturing of the eyelids).

QUESTIONS

1 Why did the lesion of the vestibulocochlear nerve give symptoms of deafness, tinnitus and giddiness?
2 What are Schwann cells?
3 What structures occupy the internal acoustic meatus?
4 Which nerve must have been damaged at surgery to account for the unilateral weakness of the facial muscles? Why was this nerve at risk?
5 Where are the facial and vestibulocochlear nerves attached to the brainstem?
6 Why was tarsorrhaphy necessary?
7 As she recovered, Mrs Maxwell had difficulty holding in her dentures when talking; why was this?

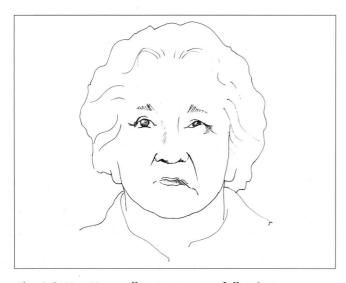

Fig. 1.2 Mrs Maxwell's appearance following treatment.

YOUR ANSWERS

1 Why the lesion of the vestibulocochlear nerve gave symptoms of deafness, tinnitus and giddiness

2 Schwann cells

3 Structures occupying the internal acoustic meatus

4 Nerve that was damaged at surgery, and reason this was at risk

5 Attachment site of facial and vestibulocochlear nerves

6 Reason tarsorrhaphy was necessary

7 Reason Mrs Maxwell had difficulty holding in her dentures when talking

See McMinn pp. 41, 132, 137, 142, 147, 169, 174, 200

Further reading
Diamond, C. and Frew, I. *The Facial Nerve*. Oxford University Press, Oxford, 1979.
Roland, P. S. and Glasscock, M. E. Acoustic neuroma. In: Paparella, M. M., Shumrick, D. A., Gluckman, J. L. and Meyerhoff, W. L. (eds.) *Otolaryngology*. Vol. II. 3rd edn. W. B. Saunders Co., Philadelphia, 1991, pp. 1775–1787.
Schwartz, J. D. *Imaging of the Temporal Bone*. Thieme Medical Publishers Inc., New York, 1986.

CASE 4

Marie-Claire Marchmont, aged 15 months, was feeding poorly, sleeping much of the time, and failing to thrive. A chest radiograph showed a mass at the root of the neck on the left side, lying adjacent to the lower cervical vertebral bodies. Products of catecholamine metabolism were found in the baby's urine, which indicated that the mass was a neuroblastoma, a malignant tumour which may arise from the sympathetic ganglia or the adrenal medulla. No metastatic spread was detected by the investigations or at surgery. The mass was excised and the diagnosis confirmed; the tumour cells were well differentiated. The surgeon gave the parents a full explanation of Marie-Claire's condition. The prognosis was favourable because of the baby's age and the early stage of the tumour.

After surgery, however, the father was particularly concerned that the child's left eyelid was drooping (ptosis). The surgeon explained that he had had to remove a section of the sympathetic trunk which supplies some of the muscle fibres which hold the eyelid open and that it would be possible to correct this later.

QUESTIONS

1 What is the extent of the sympathetic trunk?
2 Which sympathetic ganglia lie in the root of the neck?
3 Why should the neoplastic ganglion cells produce catecholamines?
4 Which muscle in the upper eyelid receives a sympathetic nerve supply?
5 What other clinical signs of sympathetic denervation might be found upon close examination?

See McMinn pp. 46–47, 162, 182. 201

Further reading
Collin, J. R. O. *A Manual of Systematic Eyelid Surgery.* 2nd edn. Churchill Livingstone, Edinburgh, 1989.
Kemshead, J. T. *Pediatric Tumours: Immunological and Molecular Markers.* CRC Press, Florida, 1989.
Mustarde, J. C. The orbital region: ptosis. In: Mustarde, J. C. and Jackson, I. T. (eds.) *Plastic Surgery in Infancy and Childhood.* 3rd edn. Churchill Livingstone, Edinburgh, 1988, pp. 136–147.
Pinkerton, C. R., Cushing, P. and Sepion, B. *Childhood Cancer Management: A Practical Handbook.* Chapman and Hall, London, 1994.
Voute, P. A., de Kraker, J. and Hoefnagel, C. A. Tumours of the sympathetic nervous system: neuroblastoma, ganglioneuroma and phaeochromocytoma. In: Voute, P. A., Barrett, A. and Lemerle, J. (eds.) *Cancer in Children: Clinical Management.* 3rd edn. Springer-Verlag, Berlin, 1992.

YOUR ANSWERS

1 Extent of sympathetic trunk

2 Sympathetic ganglia in root of neck

3 Reason for catecholamine production

4 Muscle in upper eyelid supplied by sympathetic nerve

5 Other signs of sympathetic denervation

CASE 5

Mrs Jaqueline Anderson, a 40-year-old housewife, attended the lunchtime emergency service at her local dentist, who had filled a molar tooth two days previously. She was distressed as she could no longer open her mouth and could barely speak.

The dentist recalled that he had found it difficult to anaesthetise the inferior alveolar (dental) nerve of the left side of the patient's lower jaw, for restorative treatment at the earlier appointment. He suspected that she was now suffering from trismus (an inability to open the mouth because the muscles of mastication are in spasm) due to the injection of the local anaesthetic.

Mrs Anderson was referred immediately to the dental hospital, where she was examined by a consultant oral surgeon. He confirmed the diagnosis of trismus from her history and unremarkable oral examination.

The consultant explained that local anaesthetic solution had probably been injected into the medial pterygoid muscle and that the needle had torn a blood vessel causing bleeding into the muscle (haematoma). The consequent swelling and muscle spasm prevented it from functioning normally, and thus made opening the mouth difficult. The condition would improve as the haematoma resolved. A soft diet, fluids and daily stretching exercises for the jaws were advised; Mrs Anderson was warned that the condition could last for several months.

Review appointments were made for one week and one month later, by which times there was moderate, but progressing, recovery.

QUESTIONS

1 What are the attachments and orientation of the medial pterygoid muscle?
2 For an inferior alveolar nerve block, local anaesthetic is injected close to the mandibular foramen; where is this found on the mandible?
3 From what nerve does the inferior alveolar branch, and in what region of the head does this branching occur?
4 How is the inferior alveolar nerve related to the medial and lateral pterygoid muscles on its route to the mandibular foramen?

See McMinn pp. 145–147, 152

Further reading
Allen G. D. and Hayden, J. *Complications of Sedation and Anaesthesia in Dentistry.* PSG Publishing Company Inc., Massachusetts, 1988.
Howe, G. L. and Whitehead, F. I. H. *Local Anaesthesia in Dentistry.* 3rd edn. Wright, London, 1990.
Roberts, D. H. and Sowray, J. H. *Local Analgesia in Dentistry.* 3rd edn. Wright, Oxford, 1987.

YOUR ANSWERS

1 Description of medial pterygoid muscle

2 Position of mandibular foramen on mandible

3 Nerve which gives rise to inferior alveolar nerve and region where branching occurs

4 Relationship of inferior alveolar nerve to medial and lateral pterygoid muscles

CASE 6

Mr Connor O'Rourke, a 26-year-old clerk, visited his dentist complaining of toothache on the left side of his lower jaw. A molar tooth had lost part of an amalgam filling and, from two days previously, had been throbbing, tender upon biting, and had kept him awake. His dentist offered to restore the tooth, but the patient wanted it extracted. Accordingly, the dentist administered an inferior alveolar nerve block.

To anaesthetise the inferior alveolar nerve, local anaesthetic is deposited in the pterygomandibular space, the region between the pterygoid muscles and the mandible (*Fig. 1.3*). The needle is aimed lateral to the pterygomandibular raphe from the contralateral lower premolars, in a line parallel to, and 0.5 cm above, the occlusal plane of the lower teeth.

The position of the mandibular foramen is halfway between the anterior and posterior border of the ramus; this 'depth' is gauged by holding the ramus between the thumb and forefinger of the other hand. The needle is advanced towards the foramen until bone is encountered. The needle is then withdrawn slightly, and the solution deposited smoothly, close to the foramen.

The dentist did not encounter bone with the tip of the needle but, believing that he was in the correct location, injected the anaesthetic. Three minutes later, neither the lips nor tongue felt 'frozen', and Mr O'Rourke said that the left side of his face felt strange. When asked by his dentist, Mr O'Rourke was unable to frown or close his left eye, and the left side of his mouth was drooping.

The dentist told Mr O'Rourke that the anaesthetic had spread in the wrong direction and affected the facial nerve. He said he would cover the patient's left eye with a cotton-wool patch.

The palsy arose because the local anaesthetic had been deposited beyond the pterygomandibular space, and had entered into the parotid salivary gland. This resulted in anaesthesia of the facial nerve.

Mr O'Rourke was reassured that the palsy would wear off in a few hours, and he agreed to return later that day to reattempt extraction of the tooth.

QUESTIONS

1 What is the pterygomandibular raphe and where is it situated?
2 Why did the dentist use the lower lip and tongue to assess whether the inferior alveolar nerve block was working?
3 What simple tests would confirm weakness of the muscles round the mouth?
4 Which muscle is responsible for blinking?
5 Why did the dentist cover the eye with a cotton-wool patch?
6 From your knowledge of anatomy, explain why the parotid gland had been injected and the facial nerve was anaesthetised?

YOUR ANSWERS

1 Description and site of pterygomandibular raphe

2 Reason dentist used patient's lower lip and tongue to assess nerve block

3 Tests to confirm weakness of muscles around mouth

4 Muscle for blinking

5 Reason eye covered with cotton-wool patch

6 Reason parotid gland injected and facial nerve anaesthetised

---

See McMinn pp. 144, 145–146, 152, 184

Further reading
Allen G. D. and Hayden, J. *Complications of Sedation and Anaesthesia in Dentistry.* PSG Publishing Company Inc., Massachusetts, 1988.
Howe, G. L. and Whitehead, F. I. H. *Local Anaesthesia in Dentistry.* 3rd edn. Wright, London, 1990.
Roberts, D. H. and Sowray, J. H. *Local Analgesia in Dentistry.* 3rd edn. Wright, Oxford, 1987.

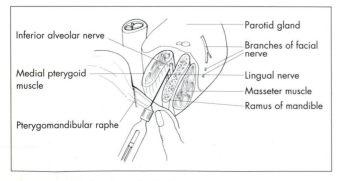

Inferior alveolar nerve

Medial pterygoid muscle

Pterygomandibular raphe

Parotid gland

Branches of facial nerve

Lingual nerve

Masseter muscle

Ramus of mandible

Fig. 1.3 Injection into the left pterygomandibular space.

CASE 7

Mr Francis McKean, a 52-year-old labourer, attended his dental surgeon in severe pain from his left upper molar teeth. The dentist found that the upper left second and third molars had fractured crowns, and thought that the remaining root portions should be extracted.

The dentist chose to anaesthetise the area using a posterior superior alveolar (dental) nerve block, since this nerve serves both these teeth.

Mr McKean opened his mouth slightly to allow placement of the needle just distally to the left upper third molar; wide opening would have caused the coronoid process of the mandible to obstruct the injection area. The needle was then advanced upwards and backwards by about 2 cm into the infratemporal region.

Immediately after deposition of the local anaesthetic solution, Mr McKean felt the left side of his face swell up. After a few minutes the swelling stopped. The dentist realised what had happened, and explained that he had probably penetrated a vein lying within that area (pterygoid plexus), causing a haematoma which might take several weeks to resolve. The complication might have been avoided by using an aspirating technique of injection.

No extractions were attempted because of the facial swelling. Mr McKean was prescribed antibiotics (in case an infection had been introduced deep into the tissues), and asked to return in one week to reattempt extraction.

QUESTIONS

1 Would the posterior superior alveolar nerve block be suitable for extraction of other upper teeth?
2 Where does the posterior superior alveolar nerve enter the maxilla?
3 What is the pterygoid plexus and where is it located?
4 Are there any special concerns about infection in the region of the pterygoid plexus?

YOUR ANSWERS

1 Suitability of posterior superior alveolar nerve block for extraction of other upper teeth

2 Entry point of posterior superior alveolar nerve

3 Description and site of pterygoid plexus

4 Concerns about infection in pterygoid plexus region

See McMinn pp. 133–134, 152

Further reading
Allen G. D. and Hayden, J. *Complications of Sedation and Anaesthesia in Dentistry*. PSG Publishing Company Inc., Massachusetts, 1988.
Howe, G. L. and Whitehead, F. I. H. *Local Anaesthesia in Dentistry*. 3rd edn. Wright, London, 1990.
Roberts, D. H. and Sowray, J. H. *Local Analgesia in Dentistry*. 3rd edn. Wright, Oxford, 1987.

CASE 8

Mr Tariq Kapoor, a 19-year-old medical student, was cycling to a clinic when he was knocked off his bike by a car suddenly emerging from a side street. He applied his brakes and swerved to avoid the vehicle but was thrown against its wing. Tariq was not badly hurt, but had abrasions across his forehead, bruises around his right eye, and was bleeding from his right nostril. The bleeding soon settled and he was taken to casualty for assessment.

On trying to take a history, however, the senior house officer (SHO) found that Tariq could not remember the accident. He could recall nothing between leaving his flat, probably about 10 minutes before the collision, and being taken to hospital. The SHO also noticed that Tariq was dabbing occasional drips of blood-tinged watery fluid from his right nostril.

A small sample of these was collected in a vial and sent for biochemical analysis. The sodium and chloride levels in the fluid lay between those for blood and tears, being lower than those expected for cerebrospinal fluid (CSF). Skull radiographs were normal.

In view of the amnesia (memory loss) of more than five minutes, however, Tariq was detained in hospital overnight for observation. Fortunately, all was well, and he was discharged the next morning.

QUESTIONS

1 A fracture of which site is suggested by black eyes and the escape of CSF into the nasal cavity?
2 For CSF to escape into the nasal cavity, what tissue layers would need to be damaged?
3 Which cranial nerves would be at particular risk in such a fracture, and how might they be tested?
4 By what route do tears reach the nasal cavity?

YOUR ANSWERS

1 Fracture site

2 Tissue layers damaged

3 Cranial nerves at risk and assessment method

4 Route of tears to nasal cavity

See McMinn pp. 130–133, 136–137, 155, 162–164

Further reading
Becker, D. P. and Gudeman, S. K. *Textbook of Head Injury.* W. B. Saunders Co., Philadelphia, 1989.
Levin, H. S., Eisenberg, H. M. and Benton, A. L. *Mild Head Injury.* Oxford University Press, Oxford, 1989.
Pitt, W. R., Thomas, S., Nixon, J. *et al.* Trends in head injuries among child bicyclists. *Br Med J* **308**, 177, 1994.
Swann, I. J. and Yates, D. W. *Management of Minor Head Injuries.* Chapman and Hall Ltd., London, 1989.

CASE 9

Annelise Brown was a three-month-old baby girl with good general health. The health visitor noticed a squint in the baby's right eye and referred her to the eye clinic. By attracting the baby's attention with a rattle and by provoking the oculocephalic reflex – in which passive rotation of the baby's head causes the eyes to move in the opposite direction, and then return to the mid-position – the orthoptist found that Annelise had a marked convergent squint on looking to the right. Inability to abduct the right eye was the only defect of ocular movement.

A full medical history revealed no family history of strabismus (squint). The obstetric history was uneventful until the second stage of labour, which had failed to progress and the cardiotocogram (fetal monitor) had indicated fetal distress. The baby was delivered using Kielland's forceps. Annelise had a good Apgar score (health assessment at birth) and the mother and baby went home three days after delivery.

It was suggested that the nerve palsy resulted from raised intracranial pressure during the difficult delivery. No other medical problems were detected and it was decided to wait up to a year to see if the nerve regenerated.

QUESTIONS

1 Which nerve had been damaged and which muscle paralysed?
2 The orthoptist used the oculocephalic reflex to make the diagnosis. Which sense organs, apart from the eyes, provide information about the position of the head?
3 Which neurological tract provides the connections between the vestibular nuclei, cervical spinal cord and the oculomotor, trochlear and abducens nuclei, so that the gaze can be fixed while the head moves?
4 Why should the abducens nerve be particularly vulnerable to raised intracranial pressure during delivery?
5 Is there any point in its course at which the abducens nerve would be vulnerable to compression by the obstetrics forceps?

YOUR ANSWERS

1 Nerve damaged and muscle paralysed

2 Other sense organs which give information on head position

3 Connecting neurological tract

4 Reason abducens nerve vulnerable to raised intracranial pressure

5 Vulnerability of abducens nerve to forceps compression

See McMinn pp. 132, 138, 167–168, 173

Further reading
de Grauw, A. J. C., Rotteveel, J. J. and Cruysberg, J. R. M. Transient sixth cranial nerve paralysis in the newborn infant. *Neuropediatrics* **14**, 164–165, 1983.

CASE 10

Mr Thomas Elder, a 52-year-old school janitor, attended the dental hospital. He had had his right upper wisdom tooth extracted three weeks previously. Removal of the molar had been accomplished without difficulty and the crown and roots were intact.

Since the operation, however, Mr Elder had noticed that, when taking fluids, drips of the drink appeared at his right nostril, particularly if his head was inclined forwards. It did not happen if he faced upwards as he drank. He also reported that, after swallowing a mouthful, a small quantity of the liquid came out of the tooth socket.

QUESTIONS

1 What is the likely diagnosis?
2 Which teeth are related to the floor of the maxillary sinus?
3 How close are the roots of the teeth to the mucosa of the sinus?
4 By what route did the fluid pass from the maxillary sinus to the anterior naris?

YOUR ANSWERS

1 Diagnosis

2 Teeth related to floor of maxillary sinus

3 Proximity of teeth roots to sinus mucosa

4 Route of fluid to anterior naris

See McMinn pp. 153–161

Further reading
Howe, G. L. *Minor Oral Surgery*. 3rd edn. Wright, Oxford, 1985.

CASE 11

Mr Des Phillips, a 20-year-old metallurgy student, was involved in a fight outside a night-club around 2 am on Christmas Eve. He was punched several times about the face.

On Christmas Day, he had a severe black eye on the left side and his left cheek was bruised and markedly swollen. He noticed that on the left side of his face, he was numb on the anterior part of the cheek, on the side of his nose and on the upper lip (*Fig. 1.4*). He reported to the accident and emergency department, where radiographic views of the facial and nasal bones showed no fractures. The swelling took about two weeks to settle, but the numbness persisted and was still present when the history was taken five months later.

QUESTIONS

1 From your knowledge of anatomy, account for the numbness on Des's face.
2 Radiology showed no fractures; which bones compose the skeleton of the face?

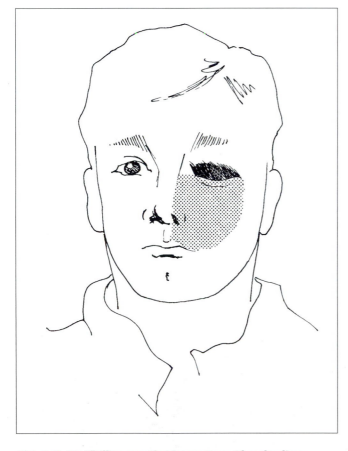

Fig. 1.4 Mr Phillips on Christmas Day. The shading shows the numb area.

YOUR ANSWERS

1 Cause of facial numbness

2 Bones of the face

See McMinn pp. 138, 142, 143, 160

Further reading
Murphy, T. M. Somatic blockade of the head and neck. In: Cousins M. J. and Bridenbaugh, P. O. (eds.) *Neural Blockade in Clinical Anaesthesia and Management of Pain.* 2nd edn. J. B. Lippincott Co., Philadelphia, 1988, pp. 533–558.

CASE 12

Mr Murray McKie, a 43-year-old animal-food sales representative, presented with a swelling in the front of his neck which he had first noticed nine months previously (*Fig. 1.5*). It had been getting gradually bigger and was now preventing him buttoning-up his shirt collars.

The lump lay near the midline and moved on swallowing. On palpation, it was firm and lay anterior to the thyroid cartilage. The mass was smooth, non-pulsatile and non-fluctuant, and measured about 2 cm × 2 cm. There were no changes in the overlying skin and no lymphadenopathy (enlarged lymph nodes). The dorsum of the tongue was inspected but no thyroid tissue was observed.

Ultrasound showed the mass to be cystic and separate from the thyroid gland. A fine-needle aspirate revealed straw-coloured fluid but no malignant cells. The diagnosis was of a thyroglossal cyst which was excised.

QUESTIONS

1 The thyroglossal cyst was a remnant of which embryonic structure?
2 Why did the mass move upwards on swallowing?
3 What is 'fluctuation'? Why was the cyst non-fluctuant, yet contained fluid?
4 Why did the surgeon look for thyroid tissue on the dorsum of the tongue?
5 What feature on the definitive tongue represents the origin of the thyroid bud? What structures, occasionally present and regarded as normal variants, are remnants of thyroid development?

Fig. 1.5 The swelling on Mr McKie's neck.

YOUR ANSWERS

1 Embryonic origin of thyroglossal cyst

2 Reason mass moved upwards

3 Fluctuation, and reason why fluid-filled cyst was non-fluctuant

4 Significance of thyroid tissue on dorsum of tongue

5 Thyroid bud origin, and thyroid remnants

See McMinn pp. 177, 180, 189–192

Further reading
Browse, N. L. *An Introduction to the Symptoms and Signs of Surgical Disease.* 2nd edn. Edward Arnold, London, 1992.

CASE 13

Mrs Nan Carson, an alert and very aware 80-year-old retired teacher, presented with a swelling in her neck. She had first noticed it two months previously and it had been slowly enlarging.

On examination, a single mass could be felt on the left side of the larynx and trachea. It was rounded and measured about 2 cm × 4 cm. It was firm, smooth, non-pulsatile and non-fluctuant. There were no changes in the overlying skin. Its upper limit was palpable but it extended through the thoracic inlet inferiorly. The mass moved upwards on swallowing.

Thyroid function tests were normal. A radio-iodine thyroid scan showed it to be part of the thyroid gland and a 'cold spot', i.e. uptake of radioactive iodine was less than that of the surroundings. Fine-needle aspiration showed the mass to be a papillary carcinoma of the thyroid gland; the prognosis for this variety of tumour is relatively good.

Mrs Carson was admitted to the general surgery ward. The left side and isthmus of the thyroid gland was removed and the left parathyroids spared. There was no lymph-node involvement.

QUESTIONS

1 What is the gross form and position of the thyroid gland?
2 Why did the mass move upwards on swallowing?
3 Why does healthy thyroid tissue take up radioactive iodine?
4 Are the thyroid and parathyroid glands intimately related when they arise in the embryo?
5 To which lymph nodes does the thyroid drain?
6 On the postoperative ward round, the consultant was careful to ensure that Mrs Carson was speaking properly; why was this?

YOUR ANSWERS

1 Form and position of thyroid gland

2 Reason mass moved upwards on swallowing

3 Reason for radioactive uptake

4 Relationship of thryoid and parathryoid glands in embryo

5 Lymph nodes draining thyroid

6 Reason for speech assessment

See McMinn pp. 188, 189–192

Further reading
Wheeler, M. H. and Lazarus, J. H. *Diseases of the Thyroid: Pathophysiology and Management.* Chapman and Hall, London, 1994.

CASE 14

This photograph of Gavin Calder (*Fig. 1.6*) was taken when he was two weeks old. There was no family history of the condition, which had not been detected by antenatal scans. The baby showed no other anomalies. Understandably, the parents were very upset.

QUESTIONS

1 What is the diagnosis?
2 From your knowledge of embryology, what failure in the development of the face has occurred to account for this appearance?
3 If the condition is not treated, what everyday functions would be particularly difficult in (a) the first year; (b) the second year onwards?

COMMENT

The care of patients with cleft lip and palate may extend from birth to maturity and depends on a dedicated specialist team, comprising a plastic and/or paediatric surgeon, orthodontist, maxillo-facial surgeon, speech therapist, audiometrician, ear, nose, and throat surgeon, specialist nurse/counsellor, geneticist, and dental specialist.

The total number of hospital visits over the years can be very large and the number of surgical procedures can vary from one for a simple cleft lip to between 10 and 20 procedures for a complicated bilateral cleft lip and palate. Each patient has individual problems but careful treatment leads to good results.

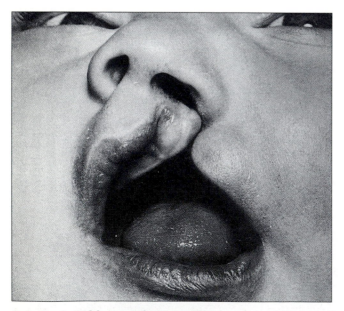

1.6 Gavin Calder, aged two weeks.

YOUR ANSWERS

1 Diagnosis

2 Developmental cause

3 Everyday functions affected in first year, and second years onwards, without treatment

Further reading
Huddart, A. G. and Ferguson, M. J. W. *Cleft Lip and Palate: Longterm Results and Future Prospects.* Manchester University Press, Manchester, 1990.
Pfeifer, G. *Craniofacial Abnormalities and Clefts of the Lip, Alveolus and Palate: Interdisciplinary Teamwork, Principles of Treatment, Longterm Results.* Thieme Medical Publishers Inc., New York, 1991.
Stengelhofen, J. *Cleft Palate: the Nature and Remediation of Communication Problems.* Churchill Livingstone, Edinburgh, 1989.

CASE 15

Miss Mary Gallacher, a recently retired lecturer in geography and tourism, had noticed over the past few months that the vision in her left eye was becoming progressively blurred; she felt as if the left side of her glasses was steamed-up, particularly when watching television. At the ophthalmology clinic, the consultant confirmed that Miss Gallacher had a cataract. After a few weeks' wait, she was admitted for replacement of the lens. Via a small incision in the superior part of the limbus, the diseased lens was excised from the lens capsule and an artificial one introduced.

Miss Gallacher has a history of glaucoma, requiring a trabeculectomy eight years ago in the left eye, and seven years ago in the right eye. She had had her eyes tested after, on a rare occasion of night driving, seeing street lamps and car-lights surrounded by hazy haloes and diagonal lines of light. The optometrist had detected increased intra-ocular pressure in both eyes and loss of peripheral vision in the right eye. Miss Gallacher had always had amblyopia in the left eye (lazy eye) with little vision.

Urgent trabeculectomy had then been carried out on the right eye; the operation involved the excision of a small piece of the sclera, iris and trabecular meshwork to allow the aqueous humour to drain freely from the eyeball. For a few months, eyedrops (pilocarpine and timolol) had been used to attempt to control the pressure in the left eye, but pressure remained high and surgery had had to be repeated.

At the current admission for cataract, the consultant was concerned that the lens surgery might cause a recurrence of glaucoma and carefully monitored the progress of healing and the intra-ocular pressure. Miss Gallacher has made an excellent recovery, however, and feels her sight is better than it has been for years.

QUESTIONS

1 Is the lens the main site of refraction in the eye?
2 What is the lens capsule and what are its attachments?
3 What is the anatomical basis of the rise in intra-ocular pressure in glaucoma?
4 What is the trabecular meshwork?
5 Pilocarpine constricts the pupil; on which muscle does it act, and is any other intra-ocular muscle likely to be affected?

YOUR ANSWERS

1 Main site of refraction

2 Lens capsule and attachments

3 Basis of rise in intra-ocular pressure in glaucoma

4 Trabecular meshwork

5 Action of pilocarpine

See McMinn pp. 164–167

Further reading
Albert, D. M. and Jakobiec, F. A. *Principles and Practice of Ophthalmology.* W. B. Saunders Co., Philadelphia, 1994.
Coakes, R. L. and Holmes Sellor, P. J. *An Outline of Ophthalmology.* Wright, Bristol, 1985.
Kanski, J. J. *Clinical Ophthalmology: a Systematic Approach.* 2nd edn. Butterworths, London, 1989.
Newell, F. *Ophthalmology: Principles and Contacts.* 7th edn. Mosby Year Book, St. Louis, 1992.

CHAPTER 2

UPPER LIMB

CASE 16

Ms Elaine Gardner, a 50-year-old care assistant, attended physiotherapy after a steroid injection into the supraspinatus muscle failed to alleviate right-sided shoulder pain. The pain had been of gradual onset over the last year.

On examination, there was 170° of shoulder flexion/elevation, with pain at the end of the range. The scapulo-humeral rhythm was normal. During abduction, a painful arc was demonstrated from 70° to 120°, after which abduction became pain-free. Resisted shoulder abduction (abduction against resistance) reproduced pain only within the first 35° of movement. All other resisted tests were negative. Palpation of the tendons of the subscapularis and infraspinatus muscles was painless. Gentle pressure on the supraspinatus tendon, however, with the patient clasping her hands behind her back, showed that the tendon was extremely tender, and caused pain to radiate down the lateral side of the arm to the elbow.

The diagnosis was supraspinatus tendinitis at the insertion. Treatment was by deep transverse frictions (a form of massage perpendicular to the line of the tendon) and ultrasound.

QUESTIONS

1 Where do the supraspinatus, infraspinatus and sub-scapularis muscles attach to the humerus? Do any other muscles attach in these regions?
2 During unresisted abduction, a painful arc was demonstrated from 70° to 120°. Abduction against resistance, however, was painful in the first 35° of movement. Are both painful episodes related to the contraction of the supraspinatus muscle?
3 When elevating the arm fully through 180°, how much movement occurs at the shoulder joint, and how much by rotation of the scapula?
4 What are the individual actions of the small muscles which attach to the greater and lesser tubercles of the humerus? Why are these muscles important?
5 Why was the pain on palpation of the supraspinatus muscle referred down the lateral side of the arm?

YOUR ANSWERS

1 Attachment to humerus of supraspinatus, infraspinatus, subscapularis, and other muscles

2 Relationship of pain to contraction of supraspinatus muscle

3 Roles of shoulder joint and scapula in arm elevation

4 Actions and importance of muscles attached to humerus

5 Reason for pain referral down lateral side of arm

See McMinn pp. 44–45, 217–218

Further reading
Apley, A. G. and Solomon, L. *Apley's System of Orthopaedics and Fractures.* 7th edn. Butterworth-Heineman, Oxford, 1993.
Kozin, F. Painful shoulder and the reflex sympathetic dystrophy syndrome. In: McCarty, D. J. (ed.) *Arthritis and Allied Conditions.* 11th edn. Lea and Febiger, Philadelphia, 1989, pp. 1509–1544.
Nicol, A. C. Biomechanics of the shoulder complex. In: Kelly, I. G. (ed.) *The Practice of Shoulder Surgery.* Butterworth-Heinemann, Oxford, 1993, pp. 17–29.

CASE 17

Mr Kevin Whitwell, a 24-year-old travel agent, was referred to the orthopaedic clinic with a two-week history of weakness at the right shoulder and difficulty raising his arm. He decided to seek medical advice when his wife noticed that his shoulder blade stuck out (winging of the scapula), particularly when he moved his arm forwards. The patient was otherwise healthy, but reported having had influenza a month previously. About a week before his wife noticed the problem, Mr Whitwell had developed a sore shoulder which had been attributed to the unaccustomed task of painting the kitchen ceiling.

The surgeon examined the shoulder and asked Mr Whitwell to push against the consulting-room wall; this confirmed winging of the right scapula (*Fig. 2.1*). Winging was not present at rest or on passive movement. The patient was sent for nerve conduction studies and electromyography, but only the long thoracic nerve was found to be involved. The diagnosis of palsy of the long thoracic nerve was explained to the patient. He was told that it might take between six months and two years to recover and that no specific treatment was required.

QUESTIONS

1 Paralysis of which muscle was suggested by winging of the scapula? Why does this sign occur?
2 Why did Mr Whitwell have difficulty raising his arm?
3 What movement of the scapula was being tested by asking the patient to push forwards against the wall?

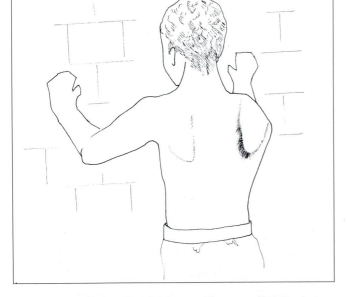

Fig. 2.1 Mr Whitwell pushing against a wall. Winging of the right scapula is seen.

YOUR ANSWERS

1 Paralysed muscle, and reason winged scapula occurs

2 Reason for difficulty in raising arm

3 Scapular movement tested by pushing forwards against wall

See McMinn pp. 213–214, 218

Further reading
Editorial. Neuralgic amyotrophy – still a clinical syndrome. *Lancet* **ii**, 729–730, 1980.
Foo, C. L. and Swann, M. Isolated paralysis of the serratus anterior: a report of 20 cases. *J Bone Joint Surg* **65B**, 552–556, 1983.
Ross, A. Neurological problems. In: Kelly, I. G. (ed.) *The Practice of Shoulder Surgery*. Butterworth-Heinemann, Oxford, 1993, pp. 269–304.

CASE 18

Mrs May O'Malley, a 62-year-old housewife, slipped on ice and stretched both hands out in front of her to break the fall. At casualty, radiographs revealed a comminuted fracture (several fragments) of the right distal radius and an undisplaced chip at the radial styloid process on the left wrist.

No stabilisation was required for the left arm. Under anaesthesia, the fractured right radius was manipulated and Kirschner wires were inserted to unite the fragments. After two weeks, the wires were removed and the right forearm was placed in plaster of Paris for five weeks with the wrist held in slight flexion.

On removal of the plaster, the right wrist lay in a degree of radial deviation, with the head of the ulna prominent. Only 40° flexion of the wrist and minimal extension was available. From the mid-prone position, a full range of pronation was present, but only 35° of supination. Accessory movements (minor gliding movements which can be produced passively by the physiotherapist but not actively by the patient) of the inferior radio-ulnar joint produced pain. In addition, the thumb was held in a degree of opposition and was very stiff at the carpometacarpal and metacarpophalangeal joints.

Treatment consisted of exercises to reduce deformity and increase the range of movement.

QUESTIONS

1 What is the radial styloid process?
2 Which bones are part of the wrist joint?
3 To which class of synovial joint does the wrist belong and what movements occur there?
4 What is the normal range of flexion and extension at the wrist?
5 At which joints do pronation and supination occur?
6 To which class of synovial joint does the carpometacarpal joint of the thumb belong, and what movements occur there?

YOUR ANSWERS

1 Radial styloid process

2 Bones of wrist joint

3 Synovial joint classification of wrist and its movements

4 Normal range of wrist flexion and extension

5 Joints at which pronation and supination occur

6 Synovial joint classification of carpometacarpal thumb joint and its movements

See McMinn pp. 230–231, 232

Further reading
Benjamin, A. Injuries of the forearm. In: Wilson, J. N. (ed.) *Watson–Jones Fractures and Joint Injuries.* 6th edn. Churchill Livingstone, Edinburgh, 1982, pp. 650–709.

CASE 19

Mrs Minnie McIntosh, an 82-year-old retired secretary, was brushing her teeth one night when her right arm lost its power – she was unable to hold her toothbrush. Over the next few minutes, the arm below the elbow became pale, cold and numb. She called her family doctor who sent her to casualty.

On examination, Mrs McIntosh's arm had become painful and was still very cold and pale. The brachial and radial pulses were absent and the skin of the forearm and hand was numb. She was in atrial fibrillation, and for the past 18 months had been taking digoxin prescribed by her family doctor. The patient was immediately treated with intravenous heparin.

The following morning, a bruit (noise heard with a stethoscope) was detected over the brachial artery, while the radial pulse was present but very weak. The patient was prepared for surgery but further assessment with Doppler ultrasound showed that the blood flow was improving spontaneously. Mrs McIntosh was started on oral warfarin and the pulses were soon restored. A cardiologist assessed the atrial fibrillation, and after four days she left hospital with her arm fully recovered.

QUESTIONS

1 What is the diagnosis?
2 Where are the brachial and radial arteries most readily palpable?
3 On initial examination, the brachial and radial pulses were absent. The forearm and hand, however, are supplied by both the ulnar and radial arteries. Would blood still be flowing in the ulnar artery in the patient?
4 What is the general layout of the arteries of the upper limb?

YOUR ANSWERS

1 Diagnosis

2 Palpation of brachial and radial arteries

3 Blood flow in ulnar artery

4 Layout of upper limb arteries

See McMinn pp. 55–56, 114, 214, 222, 226–227, 229, 231, 236

Further reading
Bennett, D. H. *Cardiac Arrhythmias.* 3rd edn. Wright, London, 1989.
Quinones-Baldrich, W. J. and Saleh, S. Acute arterial occlusion. In: Moore, W. S. (ed.) *Vascular Surgery: A Comprehensive Review.* 3rd edn. W. B. Saunders, Philadelphia, 1991, pp. 578–597.

CASE 20

Mr Walter Dryden, a 62-year-old pensioner, fell down a spiral staircase. As he fell forward, he held onto the bannister with his right arm, which was consequently pulled sharply behind him. The following day, he awoke with severe pins and needles down his right arm in, what he described as, 'showers'.

Four weeks later, Mr Dryden attended the physiotherapy department, still with severe neck and arm pain, but now also with his right arm feeling weak and shaky. He was having difficulty handling objects or carrying out fine tasks with his right hand. The feeling of pins and needles had become intermittent and occurred mainly within the area of the little and ring fingers.

On clinical testing, there was no detectable anaesthesia. There was general weakness in the movements of the fingers and thumb of the right arm. Flexion and extension were weak at the fingers and wrist. All movements of the cervical spine were painful and limited, but especially left rotation and left lateral flexion. Mr Dryden held his neck slightly flexed. On palpation, the neck was tender, but there was no localised pain to suggest a lesion of the cervical spine. Manoeuvres of the limb, designed to tense the median, ulnar and radial nerves individually, all reproduced the deep ache within the arm and severe pins and needles. There was no Horner's syndrome.

The diagnosis was a minor traction injury to the brachial plexus. Treatment was begun with pulsed short-wave diathermy followed by progressive active exercises and gentle neural stretches once the acute phase had settled.

QUESTION

From your knowledge of the sensory and motor innervation of the upper limb, at which part of the brachial plexus did the damage probably lie?

YOUR ANSWER

Area of brachial plexus damage with reasons

See McMinn pp. 42–45, 220

Further reading

Butler, D. *Mobilisation of the Nervous System*. Churchill Livingstone, Edinburgh, 1991.

Kline, D. G. and Nulsen, F. E. Acute injuries of peripheral nerves. In: Youmans, J. R. (ed.) *Youmans' Neurological Surgery*. Vol. 4. W. B. Saunders, Philadelphia, 1991, pp. 2362–2429.

CASE 21

Mrs Lorna Millar, a minister in the Church of Scotland, injured her shoulder one Sunday. She had been recovering from a painful back condition, and was anxious not to injure it again. She had therefore decided to carry her heavy bag containing her robes in front of her in both arms on the day in question. She had already conducted two morning services in her parish, and had gone in the evening to hold a service in a neighbouring parish, where the minister was on leave. After Mrs Millar had parked her car at the church, she had found the main gate closed and had to walk round to the side entrance. While doing this, her right shoulder had become painful.

Next morning, the shoulder was still painful and Mrs Millar mentioned it to a physiotherapist, whom she had met while visiting patients in the hospital. Mrs Millar said she could feel and hear the joint creaking. The pain was worst on turning her head to the side of the injury, and was felt over the shoulder and radiated into the neck. The physiotherapist examined the shoulder and found a distinct step between the acromion and the clavicle. Mrs Millar was given a sling and advised not to drive or lift anything heavy for five days.

QUESTION

From your knowledge of anatomy, what had happened to Mrs Millar's shoulder?

YOUR ANSWER

Injury to shoulder

See McMinn pp. 213, 215–216

Further reading
Apley, A. G. and Solomon, L. *Apley's System of Orthopaedics and Fractures.* 7th edn. Butterworth-Heinemann, Oxford, 1993, pp. 568–569.

CASE 22

One evening, while playing together in the bedroom, five-year-old Douglas Grant hauled his three-year-old sister, Stephanie, out of bed by her left arm. Stephanie cried out in pain, and woke up several times during that night saying her arm was sore. The next morning, Stephanie was still complaining that her arm was painful and her parents noticed it was dangling.

Stephanie's father took her to hospital where the casualty officer found that the forearm was being held in pronation, and that Stephanie cried out when he attempted to supinate it. The elbow region was tender, particularly on the lateral side, but the bony points seemed to be in their correct positions. The diagnosis was a pulled elbow in which the annular ligament had slipped over the radial head into the adjacent part of the elbow joint.

The casualty officer was able to correct this by fully supinating the forearm and then flexing the elbow; there was a slight snap as the ligament returned to its anatomical position.

QUESTIONS

1 In health, what are the attachments and position of the annular ligament?
2 To which joint does the annular ligament contribute and what movements occur there?
3 Which bony points are readily palpable in the elbow region?
4 The annular ligament was thought to have slipped over the radial head into the adjacent part of the elbow joint; between which articular surfaces would it lie?

YOUR ANSWERS

1 Attachments and position of healthy annular ligament

2 Joint to which ligament contributes, and movements of joint

3 Palpable bony points in elbow region

4 Articular surfaces between which ligament lies

See McMinn pp. 223–224, 227–228, 229

Further reading
Apley, A. G. and Solomon, L. *Apley's System of Orthopaedics and Fractures.* 7th edn. Butterworth-Heinemann, Oxford, 1993, p. 590.

CASE 23

Miss Dawn Johnstone, an 82-year-old retired nursing sister, was being visited by her niece, who had not seen her aunt for a year. During conversation, the niece noticed that her aunt's right ring and little fingers were bent. The ring finger was most affected, but both fingers were flexed, particularly at the metacarpophalangeal joint.

The niece asked about her aunt's hand and Miss Johnstone explained that the condition was called Dupuytren's contracture and that she could not extend the affected fingers. She said that it was a condition of the dense tissue within the palm, and allowed the niece to feel a firm longitudinal ridge of tissue beneath the skin of the palm proximal to the ring finger. The condition was not painful and had first appeared more than two years before, but only recently had the fingers become so markedly flexed. Her older brother had also had the same problem.

QUESTIONS

1 What is the dense fibrous tissue within the hand to which Miss Johnstone referred?
2 What is the histological structure of this dense fibrous tissue?
3 Why was there flexion of the fingers at the metacarpophalangeal joints?

YOUR ANSWERS

1 Dense fibrous tissue

2 Histological structure of dense fibrous tissue

3 Reason for flexion at metacarpophalangeal joints

See McMinn pp. 234–235

Further reading
Lister, G. *The Hand: Diagnosis and Indications.* 3rd edn. Churchill Livingstone, Edinburgh, 1993.

CASE 24

While out riding, 24-year-old Martin Robison fell off his horse when it stumbled on uneven ground. Martin stretched out his arm to break his fall and injured his left wrist, but he was otherwise unhurt. Although there was no obvious injury, his wrist was sore and made worse by attempting to move it. His friends suggested he went to the nearby casualty department.

The casualty officer found nothing on inspection, but noticed that movements were limited by pain. On palpation, tenderness was demonstrated in the anatomical snuff-box. Antero-posterior (AP), oblique and lateral radiographs of the wrist were obtained. The AP view is shown (*Fig. 2.2*).

QUESTIONS

1 What is the anatomical snuff-box?
2 Which carpal bone is palpable in the anatomical snuff-box?
3 Identify bony features A to E.
4 Can you identify the injury; what special concerns would the casualty officer have had?

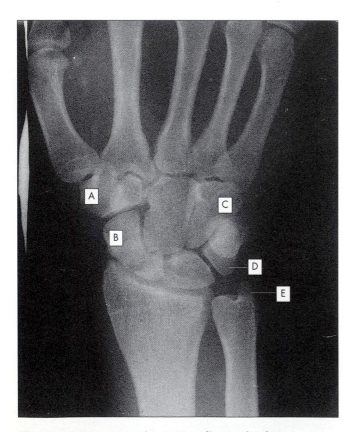

Fig. 2.2 **Antero-posterior (AP) radiograph of Mr Robison's wrist.**

YOUR ANSWERS

1 Anatomical snuff-box

2 Carpal bone palpable in anatomical snuff-box

3 Bony features A to E

4 Injury identity, and special concerns

See McMinn pp. 230–231, 236–238

Further reading
Fisk, G. R. Injuries of the wrist. In: Wilson, J. N. (ed.) *Watson–Jones Fractures and Joint Injuries*. 6th edn. Churchill Livingstone, Edinburgh, 1982, pp. 710–738.

CASE 25

Baby Fiona Goodchild was born at term (40th week of pregnancy) and had a low birthweight of 2600 g. A medical examination at 24 hours detected a continuous systolic murmur. A Doppler ultrasound scan showed a large patent ductus arteriosus. The baby was allowed to go home, but was scheduled to return at two weeks to see if the ductus arteriosus had closed spontaneously.

At the cardiac clinic, Fiona's mother reported that Fiona had been rather breathless when feeding, and she was convinced that Fiona was looking blue much of the time.

On examination, the baby appeared healthy and active. There was still a long systolic murmer with a heave (fingers on chest wall forced forwards); a thrill (palpable vibration in the chest wall corresponding to the murmur); and a low-pitched mitral diastolic murmur (due to high blood flow across the mitral valve). Ultrasound showed that the ductus arteriosus was closing slightly but was still large. The baby was admitted for observation, but the nurses reported no true cyanotic or breathless episodes; the mother, understandably, had been very anxious.

The ductus failed to close fully and was ligated when Fiona was 14 months old. Otherwise, her development was normal, and she is now a healthy toddler.

QUESTIONS

1 What is the ductus arteriosus and where is it situated?
2 In a healthy child, what happens to it after birth?
3 In fetal life, what is the direction of blood flow through the ductus?
4 The ductus arteriosus acts as a shunt to minimise blood flow through the pulmonary circulation. What is the name of the foramen in the fetal heart, which contributes to this process by allowing passage of blood from the right to the left atrium?

YOUR ANSWERS

1 Ductus arteriosus in fetus

2 Ductus arteriosus after birth

3 Direction of blood flow through ductus

4 Name of foramen between left and right atria

See McMinn pp. 111, 116, 119–120

Further reading
Gray, D. T., Fyler, D. C., Walker, A. M. *et al.* Clinical outcomes and costs of transcatheter as compared with surgical closure of patent ductus arteriosus. *N Engl J Med* **329**, 1517–1523, 1993.
Hutchison, J. H. and Cockburn, F. *Practical Paediatric Problems.* 6th edn. Lloyd-Luke, London, 1986.

CASE 26

Mr Stephen Clark, a 55-year-old accountant, was admitted to hospital because of sharp retrosternal chest pain of sudden onset on the morning of admission. He reported an episode of central chest pain when lifting a heavy box a month previously. There was a two-week history of a cough with green sputum, increasing breathlessness and associated palpitations (awareness of heartbeat).

On inspection, Mr Clark was noticed to have some swelling of the neck and difficulty in swallowing. Examination of the respiratory system revealed reduced air entry and that the trachea was central. On the right side, particularly in the lower thorax, there was:

- Decreased vocal fremitus (vibration felt by fingers on chest wall while the patient says '99').
- Hyper-resonance (hollow sound) on percussion (tapping intercostal spaces with fingers).
- On auscultation (using a stethoscope), some rhonchi (wheezes).
- Reduced vocal resonance (sound of patient speaking, heard through a stethoscope).

A chest radiograph showed a large pneumothorax (collection of air in the pleural cavity) on the right side.

After collection of sputum for bacteriology, the patient was started on antibiotics, and a chest drain, attached to an underwater seal, was inserted in the fourth intercostal space in the mid-axillary line. By the late evening, 3 l of air had been removed. The patient had mild pleuritic discomfort, however, and the drain was withdrawn.

The following morning, another chest radiograph of Mr Clark was taken. This showed the collapse of more than 50% of the right lung field. The chest drain was therefore reinserted and bubbled for three days before it was removed. The patient completed the course of antibiotics and made an uneventful recovery.

QUESTIONS

1 What is the pleural cavity and what does it contain in the healthy subject?
2 Why did Mr Clark show decreased vocal fremitus, hyper-resonance on percussion and reduced vocal resonance?
3 Through what layers would a chest drain pass when inserted at the fourth intercostal space in the mid-axillary line?
4 Why was the chest drain not inserted closer to the costal margin?
5 Would the pleuritic discomfort, experienced on the evening after admission, have arisen from the visceral or parietal pleura?
6 Can you explain the neck swelling and difficulty in swallowing?

YOUR ANSWERS

1 Description of pleural cavity and contents in healthy person

2 Reason for decreased vocal fremitus, hyper-resonance on percussion, and reduced vocal resonance

3 Layers through which chest drain passes

4 Reason why chest drain not inserted closer to costal margin

5 Origin of pleuritic discomfort

6 Reason for neck swelling and swallowing difficulty

See McMinn pp. 74–75, 100–101, 106, 120–122, 124–126, 213–214

Further reading
Compton, G. K. The respiratory system. In: Munro, J. and Edwards, C. (eds.) *Macleod's Clinical Examination.* 8th edn. Churchill Livingstone, Edinburgh, 1990, pp. 125–153.
Millard, F. J. C. and Pepper, J. R. Pleural disease. In: Brewis, R. A. L., Gibson, G. J. and Geddes, D. M. (eds.) *Respiratory Medicine.* Baillière Tindall, London, 1990, pp. 1407–1433.

CASE 27

Mrs Catriona Crawford, a 62-year-old retired caterer, underwent cardiac catheterisation for investigation of angina pectoris, with a view to possible coronary bypass surgery. The procedure involves inserting a fine catheter into the femoral artery at the groin and passing it into the ascending aorta, where contrast medium is injected into the coronary arteries. One of the radiographs obtained is shown (*Fig. 3.1*).

The patient returned to the ward about 11.50 am and 50 minutes later complained that her right leg – the side which had been catheterised – was painful. The junior house officer was called to see her. He found that Mrs Crawford's right lower limb was a dark-bluish colour and felt cold. On admission the previous day, he had recorded that the popliteal, posterior tibial, and dorsalis pedis pulses were palpable on both sides. Now, the pulses were absent on the right side. He wondered if an embolus had blocked the femoral artery and informed the registrar, who called a vascular surgeon.

It was discovered that blood had leaked out of the femoral artery at the site of the catheterisation and had compressed the artery. The surgeon corrected this and the patient made a good recovery.

QUESTIONS

1 On the angiogram, identify A, B and C.
2 The house officer tried to palpate the popliteal, posterior tibial and dorsalis pedis pulses; where may these be felt in a healthy limb?

YOUR ANSWERS

1 Identification of A, B, C

2 Positions of popliteal, posterior tibial and dorsalis pedis pulses

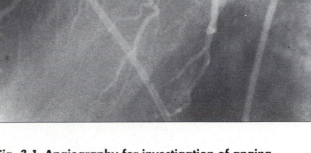

Fig. 3.1 Angiography for investigation of angina pectoris.

See McMinn pp. 114, 118, 119, 314, 318, 328, 331, 332, 335, 339

Further reading
Mendel, D. and Oldershaw, P. *A Practice of Cardiac Catheterisation.* 3rd edn. Blackwell Scientific Publications, Oxford, 1986.

CASE 28

Mrs Diana Sullivan, a 72-year-old housewife, presented to her family doctor with a history of increasing breathlessness and recent lightheadedness. She had first felt breathless about four months previously and it had become steadily worse. She now experienced breathlessness at rest and it became worse on effort. Mrs Sullivan also reported loss of weight in recent weeks. She smoked 20 cigarettes a day.

Few abnormalities were found on examination. The breathlessness at rest was confirmed. There was reduced chest expansion and enlarged lymph nodes were found in the left supraclavicular fossa. Chest radiographs were normal. Bronchoscopy, however, showed narrowing of the left principal bronchus, and a biopsy revealed the cause to be a bronchial adenocarcinoma. Computed tomography (CT) scanning showed masses in the mediastinum which were consistent with metastatic spread to the lymph nodes. A mass lay close to the left principal bronchus and another pressed on the superior vena cava. There was no evidence of a primary tumour in another organ. The tumour was inoperable, but radiotherapy alleviated the breathlessness. Mrs Sullivan was supported by the local hospice.

QUESTIONS

1 Mrs Sullivan had metastatic deposits of tumour in the regional lymph nodes of the lung. To which group of regional lymph nodes does the lung first drain?
2 Enlarged supraclavicular lymph nodes were found on clinical examination. Can you explain this finding?
3 The neoplastic cells contained mucus. What histological technique(s) would specifically stain the mucus?
4 In the healthy bronchus, how are the mucous-secreting cells arranged?
5 Enlarged bronchopulmonary lymph nodes pressed on the superior vena cava. How is the superior vena cava related to the lung root?

YOUR ANSWERS

1 Regional lymph nodes draining lung

2 Reason for enlarged supraclavicular lymph nodes

3 Histological technique(s) for staining mucus

4 Arrangement of mucous-secreting cells in healthy bronchus

5 Relationship of superior vena cava to lung root

See McMinn pp. 124, 125

Further reading

Corrin, B. The structure of the normal lungs. In: Corrin, B. (ed.) *Systematic Pathology*. Vol. 5. The Lungs. 3rd edn. Churchill Livingstone, Edinburgh, 1990, pp. 1–28.

Fishman, A. P. Cancer of the lungs. In: Fishman, A. P. (ed.) *Pulmonary Diseases and Disorders*. Vol. III. 2nd edn. McGraw-Hill Book Co., New York, 1988, pp. 1883–2066.

Haponik, E. F., Kvale, P. and Wang, K. Bronchoscopy and related procedures. In: Fishman, A. P. (ed.) *Pulmonary Diseases and Disorders*. Vol. I. 2nd edn. McGraw-Hill Book Co., New York, 1988, pp. 437–463.

Lertzman, M. Lung cancer. In: Kryger, M. H. (ed.) *Introduction to Respiratory Medicine*. 2nd edn. Churchill Livingstone, New York, 1988, pp. 277–290.

Mooi, W. J. and Addis, B. J. Carcinoma of the lung. In: Corrin, B. (ed.) *Symmer's Systematic Pathology*. Vol. 5. The Lungs. 3rd edn. Churchill Livingstone, Edinburgh, 1990, pp. 341–372.

Saunders, C. *Hospice and Palliative Care: An Interdisciplinary Approach*. Edward Arnold, London, 1990.

CASE 29

Mrs Maureen Swan is a 48-year-old solicitor with primary biliary cirrhosis awaiting a liver transplant. Her health has been poor, and over the last year she has suffered increasing chest problems and recurrent infections. Following radiographs and computed tomography (CT) scans, the diagnosis of lower lobe bronchiectasis (irreversible dilatation of the bronchi) was made by the physician. Mrs Swan was referred to physiotherapy for postural drainage and breathing exercises.

On examination, Mrs Swan was slightly jaundiced. There was no cyanosis (blue colour of the lips due to poor oxygenation of the blood) and no use of accessory respiratory muscles. On auscultation, coarse crepitations were heard in both bases, but in the left more than the right. There was also an expiratory wheeze.

Treatment was assisted by postural drainage. (This is a technique consisting of twice-daily positioning so that each bronchopulmonary segment is placed uppermost in turn. In this way, gravity is used to move secretions in the dilated bronchi towards the trachea.) Mrs Swan was asked to lie in each position for 10–15 minutes to allow optimum drainage.

During the postural drainage, Mrs Swan carried out diaphragmatic and low lateral costal breathing exercises, followed by a forced expiratory technique to facilitate expectoration of secretions.

QUESTIONS

1 Name the lobes of the right and left lungs and the fissures separating them.
2 What is the approximate surface marking of the oblique fissure?
3 Which accessory muscles of respiration might be readily seen contracting in a patient with severe respiratory problems?
4 Are the lung bases at the level of the 12th rib?
5 What is a bronchopulmonary segment?
6 How many bronchopulmonary segments are in each lobe?
7 In quiet breathing, how important is the diaphragm in both inspiration and expiration?
8 In what general direction do the lower ribs move in quiet inspiration?

YOUR ANSWERS

1 Names of the lobes of both lungs and separating fissures

2 Surface marking of oblique fissure

3 Accessory muscles in severe respiratory disease

4 Relative position of lung bases to level of 12th rib

5 Bronchopulmonary segment

6 Number of bronchopulmonary segments in lobe

7 Diaphragm in quiet breathing

8 Movement of lower ribs in quiet inspiration

See McMinn pp. 74–75, 105, 120–122, 124, 125

Further reading
Braun, S. R., Everett, E. D., Perry, M. C. and Sunderrajan, E. V. *Concise Textbook of Pulmonary Medicine.* Elsevier, New York, 1989.
Gibbons, W. J. Pulmonary rehabilitation of patients with chronic obstructive airways disease. In: Jenkinson, S. G. (ed.) *Obstructive Lung Disease.* Churchill Livingstone, New York, 1992.
Shneerson, J. *Disorders of Ventilation.* Blackwell Scientific Publications, Oxford, 1988.
Webber, B. A. *The Brompton Hospital Guide to Chest Physiotherapy.* 5th edn. Blackwell Scientific Publications, London, 1988.

CASE 30

Mrs Catherine Grierson, a 70-year-old retired postmistress, noticed a swelling in her left axilla. She did nothing about this until her left arm became swollen. There was little other history: she smoked 5–10 cigarettes per day and was taking paracetamol for osteoarthritis in her left hip. Mrs Grierson's sister had breast cancer 30 years ago.

On examination, the left upper limb was moderately swollen; this was most obvious at the wrist and hand, where pitting oedema was demonstrable. A non-tender, hard, immobile mass, about 1.5 cm in diameter, was found in the left axilla. Nothing of note was found on general examination, but an irregular, non-tender, hard mass, about 5 cm in diameter, was found in the upper medial part of the left breast. When the patient was relaxed the lump was mobile, but it became fixed on asking her to place her hands on her hips and push hard. The surgeon was careful to palpate the region between the breast and the axilla but found nothing.

At surgery, mastectomy with excision of the axillary lymph nodes was carried out. Pathological examination showed the tumour to be an infiltrating ductal carcinoma of a scirrhous-type (abundant fibrous tissue), with secondary spread to the axillary lymph nodes. Mrs Grierson received further treatment with tamoxifen.

QUESTIONS

1 Why did Mrs Grierson's arm become swollen?
2 Which muscle is made to contact by pressing the hand on the hip?
3 Why did the lump become fixed during this manoeuvre?
4 Why did the surgeon carefully palpate the region between the breast and the axilla?
5 Carcinoma of the breast often arises from neoplastic change in the duct epithelium. What is the histological appearance of the healthy duct epithelium?
6 Does the breast drain directly to any other group(s) of lymph nodes? Where are these nodes situated?
7 What is tamoxifen and how does it act?

YOUR ANSWERS

1 Reason for swollen arm

2 Muscle contracted by pressing hand on hip

3 Reason for fixation of lump

4 Reason for palpation

5 Histology of healthy duct epithelium

6 Lymph-node drainage of breast

7 Tamoxifen

See McMinn pp. 67–69, 106–107, 220

Further reading

Ahmed, A. *Diagnostic Breast Pathology: A Text and Colour Atlas.* Churchill Livingstone, Edinburgh, 1992.

Baum, M. The breast. In: Mann, C. V. and Russell, R. C. G. (eds.) *Bailey and Love's Short Practice of Surgery.* 21st edn. Chapman and Hall, London, 1992, pp. 788–821.

Blamey, R. W. The breast. In: Kyle, J., Smith, J. A. R. and Johnston, D. (eds.) *Pye's Surgical Handicraft.* 22nd edn. Butterworth-Heinemann, Oxford, 1992, pp. 334–342.

Browse, N. L. *An Introduction to the Symptoms and Signs of Surgical Disease.* 2nd edn. Edward Arnold, London, 1992.

Clain, A. *Hamilton Bailey's Demonstrations of Physical Signs in Clinical Surgery.* 17th edn. Wright, Bristol, 1986.

Rogers, K. and Coup, A. J. *Surgical Pathology of the Breast.* Wright, London, 1990.

CHAPTER 4

ABDOMEN

CASE 31

Over a period of six weeks, 55-year-old Mr Paul Whitelaw, noticed a gradually increasing swelling of his abdomen and ankles. On admission to hospital, the abdominal distension had become so severe that he felt he was going to explode. He had a history of alcohol abuse since the age of 18, and also had chronic obstructive airways disease. During the previous two years, there had been occasional episodes of frank rectal bleeding caused by haemorrhoids.

On examination, Mr Whitelaw was slightly jaundiced and had numerous spider naevi, particularly around the face and shoulders, and a slight hepatic flap; these are all signs of liver failure. The sides of his face seemed rather full, and this was attributed to enlargement of the parotid glands – a sign often associated with alcoholic liver disease.

The patient also had marked peripheral oedema. Gentle finger pressure on the skin over the ankles and sacrum produced pits which resolved over a few minutes. There was gross ascites (fluid in the peritoneal cavity), which meant that no organs were palpable. Fine crackles could be heard on auscultation of the lung bases. All these signs reflected excess tissue fluid, resulting from impaired plasma protein metabolism, as a consequence of liver failure. A sample of the ascitic fluid was obtained for analysis; it was straw-coloured, clear, and showed no bacteria or abnormal cells. Alcoholic liver disease was diagnosed.

Mr Whitelaw responded rapidly to medical treatment and special diet and fluid regimens. In four days, he lost 17 kg and his girth decreased by 20 cm. As the ascites regressed, the inferior border of the liver became palpable 10 cm below, and parallel to, the right costal margin. The tip of the spleen could also be felt just beneath the lateral part of the left costal margin. The doctor tried to discern the splenic notch, but this was not palpable. The patient was discharged home after his condition had stabilised.

QUESTIONS

1 Why was there pitting oedema over the sacrum as well as in the lower limbs?
2 The liver edge was palpable 10 cm below the costal margin; where does the inferior border of the liver lie in a healthy subject?

3 The anterior pole of the spleen was palpable through the anterior abdominal wall, just beneath the costal margin. Where does the spleen lie in a healthy subject?
4 On a healthy spleen, where is the notch located?
5 Were the episodes of rectal bleeding, recorded in the history, significant with respect to the patient's liver failure?

YOUR ANSWERS

1 Reason for pitting oedema over sacrum

2 Site of inferior border of liver in health

3 Site of healthy spleen

4 Notch location on healthy spleen

5 Significance of rectal bleeding episodes to liver failure

See McMinn pp. 266, 269–270, 273–275, 279–280, 286 (Fig. 19.35)

Further reading
Clark, M. L. and Kumar, P. J. Liver, biliary tract and pancreatic diseases. In: Kumar, P. J. and Clark, M. L. (eds.) *Clinical Medicine*. 3rd edn. Baillière Tindall, London, 1994, pp. 237–292.
Hamer-Hodges, D. W. and Munro, J. The alimentary and genitourinary system. In: Munro, J. and Edwards, C. (eds.) *Macleod's Clinical Examination*. 8th edn. Churchill Livingstone, Edinburgh, 1990, pp. 155–185.
Runyon, B. A. Care of patients with ascites. *N Engl J Med* **330**, 377–342, 1994.

CASE 32

Mrs Elizabeth Tait, a 58-year-old office cleaner, was admitted with severe pain in the upper right part of the abdomen. She had been suffering from this pain for four days, during which she had eaten little. For about a week before admission, Mrs Tait had felt nauseated and lacked appetite. She had suffered intermittent pain in the same area for several years, but the present episode was the worst and of relatively sudden onset. Previous attacks of pain had been relieved by avoidance of fatty food, but nothing stopped the present attack, which came in waves, radiated to her back, and made it difficult to lie still in bed.

At admission, the patient was noted to be well nourished. She was febrile (fevered) and her sclerae were yellow, although she was not markedly jaundiced. There was tenderness and guarding in the upper right quadrant of the anterior abdominal wall, but no masses were palpable. Laboratory and radiological tests confirmed a diagnosis of cholecystitis due to gallstones.

A cholecystectomy (gallbladder excision) was performed, and a gallstone was removed from the bile duct by a small incision in the duct wall. A T-tube was then inserted. This is a fine plastic drainage tube, shaped like a letter T, with two short limbs and one long limb. Via the incision in the bile duct, the two short limbs are inserted, one passing upwards in the duct and the other downwards. The long limb is passed through a small opening in the anterior abdominal wall. Several gallstones were found in the gallbladder.

Ten days after surgery, a T-tube cholangiogram was performed; this radiological technique checks the patency of the extrahepatic biliary apparatus by passing contrast medium into the duct system via the tube. The accompanying radiograph (*Fig. 4.1*) was obtained.

QUESTION

Can you identify features A, B, C, D, E and F? What is seen at G?

COMMENT

To remove the stone from the bile duct, the T-tube was left in place for six weeks after operation. The tube was then removed, and the stone extracted by passing a steerable catheter and stone basket along the mature track of the T-tube to the bile duct. This procedure is known as the Burhenne technique.

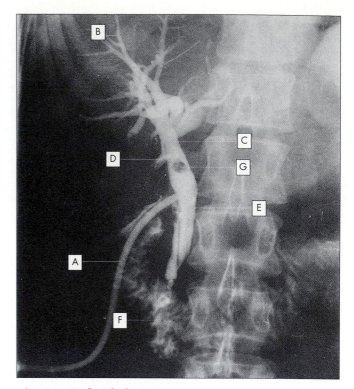

Fig. 4.1 T-tube cholangiogram.

YOUR ANSWER

Identification of features A to G

See McMinn pp. 275–277

Further reading

McMahon, A. J., Russell, I. T., Baxter, J. N. *et al.* Laparoscopic versus minilaparotomy cholecystectomy: a randomised trial. *Lancet* **343**, 135–138, 1994.

Russell, R. C. G. The gallbladder and bile ducts. In: Mann, C. V. and Russell, R. C. G. (eds.) *Bailey and Love's Short Practice of Surgery.* 21st edn. Chapman and Hall, London, 1992, pp. 1050–1076.

Yeung, E. Y. Removal of retained biliary calculi. In: *Practical Interventional Radiology of the Hepatobiliary System and Gastrointestinal Tract.* Edward Arnold, London, 1994, pp. 3–12.

CASE 33

A 49-year-old man was admitted to hospital in hypovolaemic shock (low blood pressure and rapid weak pulse due to blood loss) with haematemesis (vomiting blood), fresh bleeding per rectum, confusion, ascites (fluid in peritoneal cavity) and jaundice. He had a five months' history of alcoholic liver disease.

On admission, he was found to have a haemoglobin of 5.9 g/dl, thrombocytopenia (reduced platelets), and deranged clotting. He was resuscitated with packed red blood cells and fresh frozen plasma.

The following day, in spite of these measures, there was fresh bleeding per rectum, and his coagulopathy (deranged blood clotting) deteriorated, even after further fresh frozen plasma.

Upper gastrointestinal endoscopy revealed a substantial quantity of blood in the patient's stomach and oesophagus, as well as oesophageal varices (varicose veins in the lower oesophagus). The varices were injected and a Sengstaken tube (used to control bleeding from varices) was inserted. The patient's condition, however, continued to deteriorate, and he died the following morning.

At post-mortem, the pathologist reported the body to be that of a deeply jaundiced, middle-aged male with gynaecomastia (breast development), a markedly distended abdomen, and moderate oedema. An amount of 1800 ml of straw-coloured fluid was present in the peritoneal cavity. The mucosa of the distal end of the oesophagus contained numerous varices. Blood was found in all parts of the gastrointestinal tract from the oesophagus down to the rectum. The liver (1300 g) showed cirrhosis. The spleen (730 g) was grossly enlarged and firm; no focal abnormality was evident on sectioning.

Histology of the oesophagus showed dilated thin-walled submucosal veins indicative of oesophageal varices, while the liver showed end-stage micronodular cirrhosis, with little surviving hepatic parenchyma.

QUESTIONS

1 Were the oesophageal varices and liver failure separate conditions, or did one result from the other?
2 The patient was jaundiced, confused, had ascites, oedema and gynaecomastia, as well as impaired blood clotting. How may these clinical signs be explained?
3 Why was the spleen enlarged?
4 Is the lower oesophagus the only site of portasystemic anastomosis?

YOUR ANSWERS

1 Relationship between oesophageal varices and liver failure

2 Explanation for clinical signs

3 Reason for spleen enlargement

4 Sites of portasystemic anastomosis

See McMinn pp. 103, 273–275

Further reading
Finlayson, N. D. C., Bouchier, I. A. D. and Richmond, J. Diseases of the liver and biliary system. In: Edwards, C. R. W. and Bouchier, I. A. D. (eds.) *Davidson's Principles and Practice of Medicine.* 16th edn. Churchill Livingstone, Edinburgh, 1991, pp. 487–546.
Redmond, A. D. The 'shocked' patient: upper GI bleeding. In: Rutherford, W. H., Illingworth, R. N., Marsden, A. K., Redmond, A. D. and Wilson, D. H. (eds.) *Accident and Emergency Medicine.* 2nd edn. Churchill Livingstone, Edinburgh, 1989, pp. 143–151.
Sherlock, S. *Diseases of the Liver and Biliary System.* 8th edn. Blackwell Scientific Publications, Oxford, 1989.

CASE 34

Mr John Hannay, a 48-year-old former handyman/gardener, was admitted for investigation of dull, intermittent pain in both flanks and a mass in the left loin. He said he had been 'feeling terrible' for the past five months. Over that time, his appetite had been poor (anorexia) and he had lost 2.5 stones. He also said he felt dizzy if he stood up too quickly (postural hypotension).

Mr Hannay had been treated for hypertension for the past 22 years and, since the age of 27, had known he was suffering from polycystic kidney disease. This is a condition in which multiple cysts develop from the proximal and distal convoluted tubules of the nephrons. In some patients, cysts also occur in the liver and other organs. Mr Hannay's father, three paternal uncles, two aunts and his son had all had the disease; only the son was still alive.

Eight years ago, the patient developed gout, and he had an episode of haematuria the previous autumn. Up to then, he had smoked 30 cigarettes a day, but has cut back to 25 a week. At admission, there was no pain on micturition, urine colour was normal, and no blood was present. His hypertension and gout were controlled medically.

On examination, his blood pressure was 112/84 mmHg. The only clinical signs were in the abdomen. A large tender mass was found in the left loin. It was hard, smooth and non-pulsatile; no notch was palpable. A liver edge was palpable about 3 cm below the right costal margin. A more deeply placed mass was also palpable in the right loin inferior to the liver; it was difficult to feel and its extent vague.

Biochemical parameters indicated renal failure. Computed tomography (CT) scan showed:
- Cysts in both kidneys and in the liver.
- A renal tumour at the lower pole of the left kidney; there was no evidence of metastases.

The left kidney was removed surgically, but the right kidney function was poor and little urine was produced. Mr Hannay was therefore started on haemodialysis, and is currently awaiting renal transplantation.

QUESTIONS

1 Clearly, Mr Hannay had hereditary polycystic kidney disease; what mode of inheritance is suggested by the history?
2 What were the palpable masses in the left and right loins?
3 What substance is responsible for gout, and how might its accumulation be related to poor renal function?
4 It is not known why cysts form in patients with polycystic kidneys, although the cause may be related to abnormalities of the basement membranes of the proximal and distal convoluted tubules. Until recently, it was thought that the cysts arose from the failure of developing nephrons to connect with collecting ducts. This theory has been disproved; the continuity of cysts with nephrons has been demonstrated and cysts have been experimentally induced in rat kidneys. From which structures do the nephrons and collecting ducts develop?
5 Patients on haemodialysis usually require an arteriovenous shunt. A distended section of a vein is produced by anastomosing an artery to a vein. The distended region is readily cannulated for dialysis. The artery must have rich anastomoses with other arteries, so that tissue does not infarct (die) when the artery is attached to the vein. Which vessels might be suitable for this procedure?

YOUR ANSWERS

1 Mode of inheritance

2 Explanation of palpable masses

3 Cause of gout and link with poor renal function

4 Embryonic origin of nephrons and collecting ducts

5 Vessels suitable for arteriovenous shunt

See McMinn pp. 230, 231, 282–288

Further reading

Andrus, C. H. Vascular access. In: Jamieson, C. W. and Yao, J. S. T. (eds.) *Rob and Smith's Operative Surgery: Vascular Surgery.* Chapman and Hall Medical, London, 1994, pp. 494–505.

Baert, L. Hereditary polycystic kidney disease (adult form): a microdissection study of two cases at an early stage of the disease. *Kidney Int* **13**, 519–525, 1978.

Evan, A. P. and Gardner, K. D. Morphology of polycystic kidney disease in man and experimental models. In: Cummings, N. B. and Klahr, S. (eds.) *Chronic Renal Diseases: Causes, Complications and Treatment.* Plenum Medical Book Co., New York, 1985.

Gabow, P. A. and Grantham, J. J. Polycystic kidney disease. In: Schrier, R. W. and Gottschalk, C. W. (eds.) *Diseases of the Kidney.* 5th edn. Little, Brown and Co., Boston, 1993, pp. 535–569.

Sagger-Malik, A. K., Jeffery, S. and Patton, M. A. Autosomal dominant polycystic kidney disease. *Br Med J* **308**, 1183–1184, 1994.

CASE 35

Daniel Kent, a three-week-old baby, was being changed and bathed by his mother before an early evening feed. He became hungry and very fractious. While Daniel was crying very loudly, his mother noticed a rounded bulge at his right groin. The bulge disappeared again as he quietened down. Each time the baby cried loudly, the bulge appeared. When the family doctor was shown the site of the problem, he noted that it was just above and lateral to the scrotum. The doctor diagnosed a congenital, indirect, inguinal hernia. Daniel was admitted to the children's unit as a day patient and the hernia was surgically repaired.

QUESTIONS

1 What is an inguinal hernia?
2 Why did it appear when Daniel was crying?
3 What is the anatomical explanation for the presence of the hernia?
4 Why was the hernia 'indirect'?

YOUR ANSWERS

1 Inguinal hernia

2 Reason hernia appeared during crying

3 Anatomical reason for hernia

4 Reason hernia classed as indirect

See McMinn pp. 246–248, 308

Further reading

Hutson, J. M., Beasley, S. W. and Woodward, A. A. *Jones' Clinical Paediatric Surgery: Diagnosis and Management.* 4th edn. Blackwell Scientific Publications, Melbourne, 1992.

Spitz, L., Steiner, G. M. and Zachary, R. B. *A Colour Atlas of Paediatric Surgical Diagnosis.* Wolfe Medical Publications Ltd., London, 1981.

Tam, P. K. H. Inguinal hernia. In: Lister, J. and Irving, I. M. (eds.) *Neonatal Surgery.* 3rd edn. Butterworths, London, 1990, pp. 367–375.

CASE 36

A pathologist carried out a post-mortem on a 67-year-old woman, who had died after being in hospital for just over a month. The patient had been admitted with a two-week history of swelling of her left arm, and had also been investigated for anorexia, weight loss, diarrhoea and abdominal pain.

Tests had led to a diagnosis of malignant carcinoid tumour of the terminal ileum – a neoplasm (tumour) of entero-endocrine cells. Chest radiographs and ultrasound scans on admission had indicated the presence of metastases (secondary tumours) in the mediastinum, and thrombosis (blood clotting) in the superior vena cava, left brachiocephalic vein and left subclavian vein.

The patient was treated symptomatically and with octreotide, an analogue of somatostatin which inhibits secretion of peptides by the enteroendocrine system, and antibiotics. However, her health continued to decline steadily.

The post-mortem showed a large quantity of straw-coloured fluid (ascites) in the peritoneal cavity, and a large yellow-white polypoid (attached by stalk) tumour protruding from the mucosa of the terminal ileum. Numerous smaller tumour nodules were found on the ileum close to the primary lesion and in the small bowel mesentery. Further metastatic deposits were found around the aorta in the upper abdomen. In the mediastinum, thrombus (clot) was present in the left subclavian vein, left brachiocephalic vein and the superior vena cava. In addition, a large lobulated tumour mass, 6 cm × 5 cm × 4 cm, passed upwards from the hilar (broncho-pulmonary) lymph nodes of both sides and pressed on the superior vena cava and left brachiocephalic vein. No metastases were found in the liver.

QUESTIONS

1 What is the explanation for the presence of secondary tumour deposits in the small bowel mesentery and close to the aorta in the upper abdomen?
2 Why were no metastases present in the liver?
3 By what route were the tumour cells likely to have reached the mediastinum?
4 Where in the mediastinum do the hilar lymph nodes lie?
5 Where in the mediastinum do the left brachiocephalic vein and superior vena cava lie?

YOUR ANSWERS

1 Reason for metastases in small bowel mesentery and near to aorta in upper abdomen

2 Reason for absence of hepatic metastases

3 Route of tumour cells to mediastinum

4 Site of hilar lymph nodes in mediastinum

5 Site of left brachiocephalic vein and superior vena cava in mediastinum

See McMinn pp. 67, 69, 103, 124, 260, 273

Further reading
Dawson, I. M. P. The small intestine. In: Morson, B. C. (ed.) *Systematic Pathology.* Vol. III. Alimentary Tract. Churchill Livingstone, Edinburgh, 1987, pp. 229–291.
Sapin. M. R. and Borziak, E. I. Anatomie des ganglions lymphatiques du mediastin. *Acta Anatomica* **90**, 200–225, 1974.
Yoffey, J. M. Lymphatic system. In: Hamilton, W. J. (ed.) *Textbook of Human Anatomy.* MacMillan Press Ltd., London, 1976, pp. 279–295.

CASE 37

A post-mortem examination was carried out on the body of an 81-year-old woman. The patient had been admitted to hospital with severe melaena (black faeces due to altered blood) and required a blood transfusion. The bleeding was the result of diverticular disease of the ascending colon. Diverticular disease affects the colon and is characterised by protrusions of the colonic mucosa and submucosa through the wall, between the taeniae coli and particularly at the sites of entry of blood vessels. Problems arise when diverticulae become infected; bleeding (as has occurred in this case) is a less common presentation.

After transfusion, a right hemicolectomy (the removal of caecum, appendix and ascending colon) was performed. The terminal ileum was anastomosed to the right-hand side of the transverse colon.

Eight days after surgery, the anastomosis dehisced (broke down) and the patient was readmitted to the intensive care unit with severe sepsis and faecal peritonitis (inflammation of the peritoneum caused by spilled faeces). Further surgery was performed to reanastomose the bowel. Following the second operation, the patient's conscious level gradually deteriorated, until she died a few days later.

At autopsy, the lungs were moderately congested and oedematous. In the abdomen, the bowel anastomosis was intact and moderate diverticular disease of the sigmoid colon was noted. Pus was present in the subphrenic spaces and a large collection lay between the stomach and spleen. The liver weighed 1300 g and was moderately congested.

Histology confirmed the presence of colonic diverticula and severe inflammation on the abdominal surface of the diaphragm. Liver sections showed centrilobular necrosis, consistent with severe hepatic congestion.

QUESTIONS

1 What are the taeniae coli? Are they found all along the large intestine?
2 Why do the diverticula tend to occur between, rather than through, the taeniae coli?
3 Pus lay in the subphrenic spaces and the abdominal surface of the diaphragm was inflamed. Where are the subphrenic spaces, and why should pus have collected in them?
4 A collection of pus lay between the stomach and the spleen. In which region of the abdominal cavity does the spleen lie? Would the collection of pus be in the greater, or the lesser, sac of the peritoneal cavity?
5 The congestion of the liver probably resulted from terminal heart failure and was associated with hepatocyte necrosis around the central veins. Why were the hepatocytes in the centre of the classical liver lobule affected?

YOUR ANSWERS

1 Description and site of taeniae coli

2 Reason for diverticula between taeniae coli

3 Site of subphrenic spaces and reason for pus collection

4 Site of spleen and peritoneal site of pus collection

5 Reason for necrosis of centrilobular hepatocytes

See McMinn pp. 243–245, 264–265, 268–269, 273, 279–280

Further reading
Jass, J. R. The large intestine. In: Morson, B. C. (ed.) *Systematic Pathology.* Vol. III. Alimentary Tract. Churchill Livingstone, Edinburgh, 1987, pp. 321–325.
MacSween, R. N. M., Anthony, P. P. and Scheuer, P. J. *Pathology of the Liver.* 2nd edn. Churchill Livingstone, Edinburgh, 1987, pp. 478–502.
Mann, C. V. The small and large intestines. In: Mann, C. V. and Russell, R. C. G. (eds.) *Bailey and Love's Short Practice of Surgery.* 21st edn. Chapmann and Hall, London, 1992, pp. 1125–1167.

CASE 38

Pauline Charteris, a 14-year-old schoolgirl, was admitted to the acute surgery ward with a very sore abdomen. When it had started the previous evening, the pain was colicky, intermittent, and was felt around the umbilicus. The following morning, the pain had moved to the right iliac fossa and the family doctor was called. By the time he had arrived 30 minutes later, the pain was worse, constant, and intensified by moving. Pauline had vomited through the night and had had two episodes of diarrhoea. She had not eaten for two days. Her face was flushed, her tongue was furred, and her breath was bad.

On examination, the abdomen was very tender all over, and there was guarding (tensing of the abdominal muscles on palpation of the tender area). Pauline showed rebound tenderness (increased pain when the doctor lifted his hand from the abdomen) all over the abdomen. Pressure on the left lower quadrant caused pain in the right lower quadrant (Rovsing's sign). Maximum tenderness was over McBurney's point.

On admission, Pauline was taken urgently to theatre for an appendicectomy. The operation was performed through a gridiron incision (see below). On opening the abdomen, the caecum was seen and the appendix was soon located. The appendix was found to be ruptured with a faecolith (hard faeces) in its lumen. The abdominal cavity was lavaged with 3 l of saline at 37°C containing antibiotics.

Pauline remained in hospital for a further three days and received intravenous antibiotics and analgesia before going home.

QUESTIONS

1 Why was pain from the appendix initially felt in the umbilical region?
2 Why did the pain move to the right iliac fossa as the inflammation progressed?
3 Why did the character of the pain change as it moved to the right iliac fossa?
4 Where is McBurney's point and what is its significance?
5 What important layers would be divided on making an incision in the region of McBurney's point?
6 The gridiron technique involves making incisions in each muscular layer parallel to the line of its fibres. In approximately what directions do the fibres of these layers run at McBurney's point?
7 Are the fibres of the external oblique muscle fleshy or aponeurotic at McBurney's point?
8 What nerve is at particular risk using this approach? What might be the long-term effect of damaging this nerve?
9 What features of the caecum are used to locate the appendix?

YOUR ANSWERS

1 Reason for pain in umbilical region

2 Reason pain moved to right iliac fossa

3 Change in character of pain

4 McBurney's point

5 Layers divided by incision at McBurney's point

6 Directions of muscle-layer fibres at McBurney's point

7 Character of external oblique muscle fibres at McBurney's point

8 Nerve at risk and long-term effects of damage

9 Features of caecum

See McMinn pp. 243, 246–248, 263–264, 268–269

Further reading
Browse, N. L. *An Introduction to the Symptoms and Signs of Surgical Disease.* 2nd edn. Edward Arnold, London, 1992.
Clain, A. *Hamilton Bailey's Demonstrations of Physical Signs in Clinical Surgery.* 17th edn. Wright, Bristol, 1986.
Forrest, A. P. M., Carter, D. C. and Macleod, I. B. *Principles and Practice of Surgery.* 2nd edn. Churchill Livingstone, Edinburgh, 1991.
Newstead, G. C. Open appendicectomy. In: Fielding, L. P. and Goldberg, S. M. (eds.) *Operative Surgery: Surgery of the Colon, Rectum and Anus.* 5th edn. Butterworth-Heinemann Ltd., Oxford, 1993, pp. 337–346.

CASE 39

Mr Matthew Beattie, a 74-year-old retired lighthouse keeper, was admitted to hospital when he became jaundiced over a six-day period. He had lost weight in recent weeks but had been previously well. In the few days prior to admission, he had noticed pale faeces, difficult to flush away. On examination, there were few positive findings, but the gall bladder was palpable. The inferior border of the liver was also palpable about 1 cm below the costal margin.

In hospital, a number of investigations were carried out. Serum biochemistry suggested the jaundice was obstructive in nature. Abdominal ultrasound confirmed the diagnosis of carcinoma of the head of the pancreas. Mr Beattie was fully informed about his condition and agreed to the operation of cholecystjejunostomy, in which the distended gall bladder was anastomosed to the jejunum. This provided a palliative short-circuit which prevented the discomfort and pruritis (itch) associated with obstructive jaundice.

Now that the operation is over, Mr Beattie has been able to return home with his symptoms gone. He is attending the hospice in his neighbourhood, where residential care will be available as the disease progresses.

QUESTIONS

1 What is the surface marking of the fundus of the gall bladder?
2 Why did the carcinoma of the head of the pancreas result in jaundice?
3 Following cholecystjejunostomy, by what route does the bile enter the intestine?

YOUR ANSWERS

1 Surface marking of gall-bladder fundus

2 Reason pancreatic carcinoma caused jaundice

3 Route of bile after cholecystjejunostomy

See McMinn pp. 275–277, 277–279

Further reading
Ashley, S. W. and Reber, H. A. Surgical management of exocrine pancreatic cancer. In: Go, V. L. W. *et al.* (eds.) *The Pancreas: Biology, Pathobiology and Disease.* Raven Press Ltd., New York, 1993, pp. 913–929.
Mills, M., Davies, H. T. O. and Macrae, W. A. Care of patients dying in hospital. *Br Med J* **309**, 583–586, 1994.
Preece, P. E., Cuschieri, A. and Rosin, R. D. *Cancer of the Bile Ducts and Pancreas.* W. B. Saunders Co., Philadelphia, 1989.
Russell, R. C. G. The pancreas. In: Mann, C. V. and Russell, R. C. G. (eds.) *Bailey and Love's Short Practice of Surgery.* 21st edn. Chapman and Hall, London, 1992, pp. 1077–1099.

CHAPTER 5

PELVIS

CASE 40

Mr Jan Wozna, a 66-year-old retired accountant, had a 2–3-year history of difficulty in initiating micturition. In recent months, frequency of micturition was becoming troublesome, and he went to the toilet usually three times a night. The flow seemed much poorer than it had been when he was younger. Mr Wozna knew that these were symptoms of an enlarged prostate, and he was concerned because his old boss and golfing partner had suffered acute urinary retention the year before.

General examination showed Mr Wozna to be in good health, but the rectal examination revealed an enlarged prostate. The prostate was symmetrical, with a rubbery consistency. A vertical midline groove was palpable.

Radiographic and ultrasound examination of the bladder and ureters showed no specific defects. Normal serum levels of acid phosphatase and prostate-specific antigen provided reassurance that there was no prostate cancer.

Mr Wozna was treated by transurethral resection of the prostate, a procedure in which diseased tissue is removed, slice by slice, through a resectoscope passed along the urethra.

The adenoma (benign growth) usually affects the cranial part of the prostate and the compressed outer caudal tissue, seminal colliculus (verumontanum) and urethral sphincter, are retained. The patient was warned about the possibility of retrograde ejaculation (in which semen enters the bladder on ejaculation).

For 48 hours following the operation, a urinary catheter was in place, and Mr Wozna was somewhat alarmed by blood in the urine. He was reassured and the blood ceased by the time the catheter was removed.

QUESTIONS

1. Through which structures does the male urethra pass, and at which part of its course does it pass through the prostate?
2. What is the seminal colliculus and what are its features?
3. Where do the ejaculatory ducts commence and how do they reach the prostatic urethra?
4. What is the histological structure of the prostate?
5. Blood clots may be passed for a day or two after transurethral resection; has this blood haemorrhaged from the prostatic venous plexus?

YOUR ANSWERS

1 Structures male urethra passes through

2 Seminal colliculus

3 Ejaculatory ducts and route to prostatic urethra

4 Histology of prostate

5 Source of postoperative haemorrhage

See McMinn pp. 295, 308–309, 311

Further reading

Blandy, J. *Lecture Notes on Urology.* 4th edn. Blackwell Scientific Publications, Oxford, 1989.

Gosling, J. A., Dixon, J. S. and Humpherson, J. R. *Functional Anatomy of the Urinary Tract.* Churchill Livingstone, Edinburgh, 1983.

Kirk, D. How should new treatments for benign prostatic hyperplasia be assessed? *Br Med J* **306**, 756–760, 1993.

CASE 41

Ms Gillian Waugh, a 24-year-old, nulliparous, primary-school teacher, had been trying to become pregnant for almost two years, and was distressed by her lack of success. Her family doctor noted that she was a non-smoker, who menstruated regularly for four days every 30 days, and who gave no history of pelvic inflammatory disease or abdominal surgery. Examination failed to reveal any abnormality, while her hormonal profile strongly suggested the occurrence of ovulation and was otherwise unremarkable. Ms Waugh's partner's semen sample was satisfactory and the couple did not admit to any sexual dysfunction.

Suspecting a tubal disorder, the family doctor referred Ms Waugh for further investigation. A laparoscopic exam-ination demonstrated multiple pelvic adhesions, possibly the result of a clinically silent infection. Cervical injection of methylene blue dye (hydrotubation) was unable to confirm the patency of either uterine tube. A subsequent hystero-salpingogram (radiological visualisation of the genital tract) confirmed the occlusion of both tubes and, curiously, showed an arcuate uterus with septum (*Fig. 5.1*). This congenital abnormality was corrected by a plastic operation, and the couple later conceived with the assistance of an *in vitro* fertilisation (IVF) technique.

QUESTIONS

From your knowledge of embryology, can you explain the appearance of the uterus on the hysterosalpingogram?

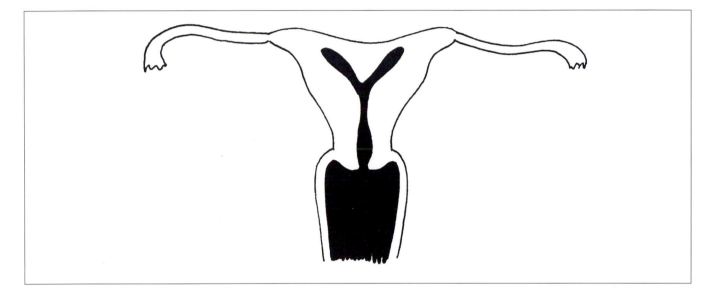

Fig. 5.1 **Arcuate uterus with septum.**

See McMinn pp. 299–300

Further reading

Asmo, N. N. and Shaw, R. W. New frontiers in assisted reproduction. In: Shaw, R., Soutter, P. and Stanton, S. (eds.) *Gynaecology*. Churchill Livingstone, Edinburgh, 1992, pp. 231–248.

Healy, D. L., Trounson, A. O. and Andersen, A. N. Female infertility: causes and treatment. *Lancet* **343**, 1539–1544, 1994.

Sadler, T. W. *Langman's Medical Embryology*. 6th edn. Williams and Wilkins, Baltimore, 1990.

Thompson, W. and Heasley, R. N. Investigation of the infertile couple. In: Shaw, R., Soutter, P. and Stanton, S. (eds.) *Gynaecology*. Churchill Livingstone, Edinburgh, 1992, pp. 219–229.

Tindall, V. R. *Jeffcoate's Principles of Gynaecology*. 5th edn. Butterworths, London, 1987.

YOUR ANSWER

Reason for appearance of uterus

CASE 42

Mrs Nance Boyd, a 54-year-old widow and mother of four, presented to her family doctor complaining of a sensation of 'something coming down' her vagina when she was on her feet, and backache which was worse at the end of the day. Mrs Boyd was also troubled by stress incontinence whenever she coughed, sneezed, or carried out routine activities, such as making her bed or carrying the groceries home. She also described having to micturate about 10 times during the day (frequency) and five times at night (nocturia). She had reached the menopause at 51 years of age.

Pelvic examination revealed vulval ulceration, first-degree uterovaginal prolapse (cervix still inside the vagina), and a prominent cystocele (prolapse of the bladder base) (*Fig. 5.2*). The latter formed an obvious bulge of the upper anterior vaginal wall. Mrs Boyd reluctantly admitted that she had to apply digital pressure to the bulge to empty her bladder completely.

Mrs Boyd's doctor advised her to stop smoking and to lose weight, and referred her to a gynaecologist who recommended a Manchester repair. This operation involves the shortening of the transverse cervical ligaments and amputation of the cervix.

It is done together with an anterior colporrhaphy, in which redundant anterior vaginal wall is removed and the pubovesical fascia is approximated. Patients in whom there is (or is likely to be) a prolapse of the posterior vaginal wall require posterior colporrhaphy, but this was not necessary in Mrs Boyd's operation.

Mrs Boyd made an uneventful recovery from the surgery and was cured of her distressing symptoms.

QUESTIONS

1 What sphincter controls urinary flow?
2 Why do sneezing, coughing and lifting produce escape of urine in patients with stress incontinence?
3 Where do the transverse cervical ligaments lie and why are they shortened during the Manchester repair?
4 The cystocele caused a bulge in the upper part of the anterior vaginal wall; what lies anterior to the lower part of the wall?
5 What structures could press against, and cause prolapse of, a weak posterior vaginal wall?

YOUR ANSWERS

1 Sphincters controlling urinary flow

2 Reason for stress incontinence

3 Site of transverse cervical ligaments and reason for Manchester repair

4 Structure anterior to lower cystocele wall

5 Structures causing vaginal prolapse

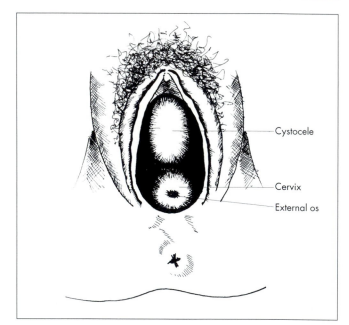

Fig. 5.2 Cystocele seen on vaginal examination.

Cystocele
Cervix
External os

See McMinn pp. 288–289, 290–291, 294–296, 299, 303 (Fig. 19.44)

Further reading
Lewis, T. L. T. and Chamberlain, G. V. P. *Gynaecology by Ten Teachers*. 15th edn. Edward Arnold, London, 1990.
Stanton, S. L. Vaginal prolapse. In: Shaw, R., Soutter, P. and Stanton, S. (eds.) *Gynaecology*. Churchill Livingstone, Edinburgh, 1992, pp. 437–447.
Willocks, J. and Neilson, J. P. *Student Notes: Obstetrics and Gynaecology*. 4th edn. Churchill Livingstone, Edinburgh, 1991.

CASE 43

Ms Amy McDonald, a 33-year-old bank clerk, presented to her family doctor with a long history of painful, heavy periods. The pain (dysmenorrhoea) was localised to her lower abdomen; it began premenstrually, reached a peak during menstruation, and subsided slowly thereafter. Ms McDonald menstruated regularly for seven days every 28 days. She would often pass large clots together with, in her opinion, an excessive amount of blood (menorrhagia). She also complained of fatigue and, on careful questioning, admitted to being troubled by pelvic pain during sexual intercourse (deep dyspareunia).

This pain was reproduced during bimanual examination of the pelvis, when the vaginal hand palpated a mass in the recto-uterine pouch (pouch of Douglas), while the abdominal hand detected the presence of the uterus in the expected place, thereby suggesting the presence of a retroverted uterus. The family doctor suspected endometriosis, a common gynaecological lesion consisting of ectopic deposits of endometrium in the pelvic cavity, and referred Ms McDonald to a specialist for investigation.

Laparoscopy confirmed the family doctor's clinical findings, with widespread endometriosis observed throughout the lower part of the peritoneal cavity and a retroverted uterus fixed by multiple adhesions. The gynaecologist vapourised the endometrial deposits and adhesions with a laser, and prescribed a six-month course of danazol, a synthetic steroid closely related to testosterone. This relieved Ms McDonald of her symptoms.

QUESTIONS

1 The uterus was palpated on bimanual examination; may it be felt by palpating the anterior abdominal wall only?
2 What is meant by the term, 'retroverted uterus'?
3 Why did Ms McDonald's pain occur around the time of her period?
4 Would bimanual examination have been appropriate had Ms McDonald been a virgin?

YOUR ANSWERS

1 Uterus palpation

2 'Retroverted uterus'

3 Reason for pain around menstruation

4 Appropriateness of examination in virgin

See McMinn pp. 299–301, 301–303

Further reading

Shaw, R. W. Endometriosis. In: Shaw, R., Soutter, P. and Stanton, S. (eds.) *Gynaecology.* Churchill Livingstone, Edinburgh, 1992, pp. 421–435.

Thomas, E. J. Endometriosis. *Br Med J* **306**, 158–159, 1993.

Tindall, V. R. *Jeffcoate's Principles of Gynaecology.* 5th edn. Butterworths, London, 1987.

CASE 44

Mrs Lucia Jackson, a 42-year-old Jamaican woman, presented to the accident and emergency department of her local hospital late one evening, with a short history of stabbing lower abdominal pain, shoulder-tip pain on both sides, and pain on defaecation. She was bleeding from her genital tract, but the blood was dark brown and scanty, quite unlike her normal menstrual flow. Further questioning revealed that, although Mrs Jackson usually menstruated regularly for three days every 28 days, her last menstrual period had been more than eight weeks previously. This absence of menstruation (amenorrhoea) had caused her little concern, being attributed to the 'change of life'.

The abdomen showed tenderness and guarding in its lower parts. Vaginal examination was not performed, in case the diagnosis was ectopic pregnancy with consequent risk of sudden haemorrhage.

Mrs Jackson explained to the casualty officer that she had been sterilised several years previously and therefore could not be pregnant. However, to her surprise, a serum beta-human chorionic gonadotrophin (beta-HCG) test indicated that she was indeed pregnant. Following an ultrasound scan, which showed an empty, enlarged uterus with free fluid in the recto-uterine pouch, a tentative diagnosis of ectopic pregnancy was made. At laparoscopy, blood was found in the peritoneal cavity and the ampulla of the right uterine tube was swollen, consistent with the diagnosis. Salpingectomy (removal of the affected tube) was performed.

QUESTIONS

1 Where in the pelvis does the uterine tube lie?
2 Name the regions of the uterine tube.
3 In which region does fertilisation usually take place?
4 Which organs usually lie in the recto-uterine pouch?
5 How is sterilisation accomplished?
6 Why did Mrs Jackson have shoulder-tip pain?

YOUR ANSWERS

1 Site of uterine tube

2 Regions of uterine tube

3 Usual site of fertilisation

4 Organs in recto-uterine pouch

5 Sterilisation procedure

6 Reason for shoulder-tip pain

See McMinn pp. 101–103, 210–211, 299, 305, 307

Further reading

Cacciatore, B., Sterman, U. and Ylostalo, P. Early screening for ectopic pregnancy in high risk, symptom free women. *Lancet* **343**, 517–518, 1994.

Elias, J. A. Sterilization. In: Loudon, N. (ed.) *Handbook of Family Planning.* 2nd edn. Churchill Livingstone, Edinburgh, 1991, pp. 243–267.

Margara, R. A. Ectopic pregnancy. In: Shaw, R., Soutter, P. and Stanton, S. (eds.) *Gynaecology.* Churchill Livingstone, Edinburgh, 1992, pp. 279–290.

CASE 45

Christine and Colin Ogilvie are a newly married couple who are keen to have a family. A few months ago, Christine, a 26-year-old shop assistant, suspected she was pregnant. It was nine weeks since her last menstrual period; she noticed a fullness in her breasts, and for two weeks she had been troubled by nausea and vomiting. A urine test confirmed the pregnancy, and her family doctor carried out a general examination. The uterus was not yet palpable on abdominal examination.

The next week, Christine and Colin attended the local maternity hospital for an ultrasound scan. It indicated that the fetus was in the eighth week of development. The gestational age is determined by the crown rump length; at this stage, details of the fetus are not yet visible.

At a later visit to the doctor, at week 17 of the pregnancy, a sample of blood was sent for alpha-fetoprotein (AFP) analysis. The following afternoon, the laboratory phoned the doctor to say that the levels were abnormally high.

Two days later, the couple attended the hospital for a further ultrasound scan. The radiologist found that the usual spherical outline of the head was absent and that the skull vault was defective; this is diagnostic of anencephaly, a severe defect of brain formation caused by failure of the anterior neuropore to close. The obstetrician counselled the couple, who agreed to have the pregnancy terminated. This was carried out two days later; the fetus had a flat head, no skull vault, and an exposed mass of vascular tissue. The fetus also showed an omphalocele, which is a herniation of loops of midgut into the umbilical cord.

The couple were advised to try for another pregnancy, but that there was a 3–5% risk that the next baby might also have a neural tube defect. Four months after the termination, Christine became pregnant again; this time all is well, and the couple are looking forward to the birth.

QUESTIONS

1 What substance is detected in the urine test for pregnancy, and what is its source?
2 When in pregnancy does the uterus become palpable on abdominal examination?
3 Why does analysis of AFP help in the detection of neural tube defects?
4 At what stage of development do the anterior and posterior neuropores close?
5 What condition results if the posterior neuropore fails to close?
6 At what stage do loops of midgut usually lie in the umbilical cord?

YOUR ANSWERS

1 Substance in urine and its source

2 Stage that uterus is palpable

3 AFP and neural tube defects

4 Stage at which neuropores close

5 Condition due to failed closure of posterior neuropore

6 Stage at which midgut loops found in umbilical cord

See McMinn pp. 304–305

Further reading
Brock, D. J. H. Mechanisms by which amniotic fluid alpha-fetoprotein may be increased in fetal abnormalities. *Lancet* ii, 345–346, 1976.
de Crespigny, L. and Dredge, R. *Which Tests for My Unborn Baby? A Guide to Prenatal Diagnosis.* Oxford University Press, Melbourne, 1991.
Neilson, J. P. and Chambers, S. E. *Obstetric Ultrasound.* Oxford University Press, Oxford, 1993.
Whittle, M. J. and Connor, J. M. *Prenatal Diagnosis in Obstetric Practice.* Blackwell Scientific Publications, Oxford, 1989.

CHAPTER 6

LOWER LIMB

CASE 46

Ms Henrietta Rowan, an 18-year-old student, went skiing in Austria in the winter of 1991–92. While on the slopes on the first day, she collided with a 14-year-old boy, who landed on the outside of Henrietta's right knee. The knee was very sore; Henrietta was unable to move and had to be airlifted by helicopter to the nearest hospital, just over the border into Germany.

At the casualty department, the receiving doctor examined Henrietta's knee, which was swollen but not particularly bruised. He found that a small amount of abduction was permissible at the knee, and that if he clasped his hands behind Henrietta's calf and pulled anteriorly, there was some forward movement of the tibia relative to the femur. The testing procedure was painful for the patient. A diagnosis of ligamentous damage at the knee joint was made and a temporary plaster was applied. Henrietta was then flown back to her family home in London.

The next day, Henrietta was seen at the orthopaedic clinic of the local hospital, where further tests confirmed partial damage to two ligaments (see below). The plaster was revised to extend from the mid-thigh to the ankle and was kept on for eight weeks. For the first two weeks, Henrietta had to walk on crutches, after which she began to start using the limb again for walking. Ibuprofen was prescribed for pain.

When the plaster was removed, there was much muscle wasting, particularly noticeable in the quadriceps. A three-month course of physiotherapy was prescribed.

Two years later, the knee has not fully recovered. Henrietta still feels weakness in the joint and complains that it sometimes 'gives' without warning during weight-bearing.

See McMinn pp. 315, 324–327

Further reading
Abernethy, P. J. The locomotor system. In: Munro, J. and Edwards, C. (eds.) *Macleod's Clinical Examination.* 8th edn. Churchill Livingstone, Edinburgh, 1990, pp. 255–296.
Apley, A. G. and Solomon, L. *Apley's System of Orthopaedics and Fractures.* 7th edn. Butterworth-Heinemann, Oxford, 1993.
Stanish, W. D., O'Grady, P. and Dillon, J. E. Knee ligament sprains – acute and chronic. In: Harries, M., Williams, C., Stanish, W. D. and Micheli, L. J. (eds.) *Oxford Textbook of Sports Medicine.* Oxford Medical Publications, New York, 1994, pp. 364–384.

QUESTIONS

1 The receiving doctor at the German casualty department found excessive abduction at the knee; damage to which ligament is suggested by this observation?
2 Damage to which ligament is suggested by the excessive forward movement of the tibia relative to the femur?
3 Which individual muscles compose the quadriceps femoris and what is their action on the knee?
4 Can you offer any explanation for why the knee might give way during weight-bearing?

YOUR ANSWERS

1 Ligament damage suggested by excessive abduction

2 Ligament damage suggested by excessive forward movement of tibia

3 Muscles comprising quadriceps femoris and action on knee

4 Reason why knee gave during weight-bearing

CASE 47

Neil Ferguson, a 55-year-old electrical engineer, is to have a total replacement of his right knee. During the past three years, he has had increasing pain in his knee. The pain is worse on walking, which is limited to less than a mile per day. There is no morning stiffness. He has also become slightly bow-legged.

At the age of 18, he twisted his right knee while playing rugby. The knee had been painful and slightly swollen for a few days. Over the next two years, he had been troubled by his right knee being intermittently painful and occasionally 'giving way'. Sometimes the knee locked, but could be 'clicked back'. A tear of the medial meniscus was diagnosed and medial meniscectomy was performed when Mr Ferguson was aged 20.

At the age of 45, the patient's knee became swollen and painful. It felt unstable, especially on extension. This was treated by physiotherapy in the form of quadriceps exercises and the condition settled.

On examination during the present admission, varus (adduction) deformity of the right knee was noted. The old meniscectomy scar was noted medial to the patella. The knee region was tender on palpation, especially at the medial side, but there was no effusion. Radiography showed reduced joint space of the medial compartment with osteophyte formation; clearly, the joint had been severely affected by osteoarthritis.

QUESTIONS

1 Mr Ferguson had torn his right medial meniscus playing rugby as a young man; which meniscus is more commonly damaged and why?
2 Of what is the medial meniscus composed and what are its shape and attachments?
3 Why should the knee have sometimes locked and been able to be 'clicked back'?
4 What is the relationship of the quadriceps to the patella, and how is this group of muscles attached to the tibia?
5 The joint space of the medial compartment of the knee was reduced. What is meant by the term 'medial compartment', and what accounts for the radiological appearance of a space between the bones?

YOUR ANSWERS

1 Meniscus most commonly damaged and reason why

2 Description of medial meniscus

3 Reason knee locked and 'clicked back'

4 Relationship of quadriceps to patella and attachment to tibia

5 Meaning of 'medial compartment' and cause of radiological space

See McMinn pp. 315, 324–327

Further reading

Apley, A. G. and Solomon, L. *Apley's System of Orthopaedics and Fractures.* 7th edn. Butterworth-Heinemann, Oxford, 1993.

Brady, O. H. and Hutson, B. J. Acute injuries of the meniscus. In: Harries, M., Williams, C., Stanish, W. D. and Micheli, L. J. (eds.) *Oxford Textbook of Sports Medicine.* Oxford Medical Publications, New York, 1994, pp. 350–363.

CASE 48

Ms Laura Taylor, aged 23, presented with a recurring deep venous thrombosis in her right leg. The limb was red, tender and swollen from the groin to the foot and felt hot. She had been in hospital with a similar episode two months previously and was treated with warfarin, but the symptoms had returned in the two days prior to admission.

Laura is an intravenous drug abuser, injecting temazepam and heroin into her groin. She had been injecting herself for five years, and the veins in her groin were the only accessible patent veins. The money for the drugs comes largely from prostitution.

An ultrasound scan showed clots in Laura's popliteal and femoral veins, but they were not totally occluded. After initial treatment with heparin, she was prescribed oral warfarin and, when the symptoms settled, she was discharged into the care of her family doctor. Laura would have to stop injecting into her groin before the thrombosis could be adequately cured.

QUESTIONS

1 Where, in terms of surface anatomy, does the femoral vein lie at the groin?
2 Which veins in the lower limb are readily accessible for injection?
3 What is the general arrangement of the deep veins of the lower limb?
4 What factors facilitate the return of blood from the deep veins of the lower limb?

YOUR ANSWERS

1 Site of femoral vein at groin

2 Lower limb veins accessible for injection

3 Arrangement of deep veins of lower limb

4 Factors facilitating return of blood

See McMinn pp. 314–315, 317, 318, 332

Further reading
de Bono, D. P. and Boon, N. A. Diseases of the cardiovascular system. In: Edwards, C. R. W. and Bouchier, I. A. D. (eds.) *Davidson's Principles and Practice of Medicine.* 16th edn. Churchill Livingstone, Edinburgh, 1991, pp. 249–339.
Drug Misuse and Health in Glasgow. Greater Glasgow Health Board, 1993.
Greenfield, L. J. Venous thromboembolic disease. In: Moore, W. S. (ed.) *Vascular Surgery: a Comprehensive Review.* 3rd edn. W. B. Saunders, Philadelphia, 1991, pp. 669–679.
Griffen, S., Peters, A. and Reid, M. Drug misusers in Lothian: changes in injecting habits 1988–90. *Br Med J* **306**, 693, 1993.
Plant, M. A. *Aids, Drugs and Prostitution.* Tavistock Routledge, London, 1991.

CASE 49

Mr Gerry Lucas, a 36-year-old physical education teacher, was referred for physiotherapy with pain in the region of the right heel. He had suffered similar problems with his left heel 18 months previously, which had resolved with physiotherapy and the use of a foot orthosis (special footwear) provided by the podiatrist, who had diagnosed flat feet.

The right-heel pain had been intermittent over the last year, but was now constant. Pain was increased by walking and running. There was an effusion around the lower medial border of the tendo calcaneus (Achilles tendon) and pain on palpation within this area. Resisted plantar flexion reproduced the pain. During gait, pain was most significant during the push-off element of the stance phase. Radiology showed no abnormality.

Treatment of this Achilles tendonitis consisted of deep transverse frictions (vigorous massage perpendicular to the line of the tendon to prevent adhesion to surrounding structures and increase the circulation), ultrasound, advice on footwear and referral to the podiatrist for an orthosis.

QUESTIONS

1 Which muscles insert via the tendo calcaneus?
2 Why is this muscle group sometimes called the triceps surae?
3 An effusion was noted at the lower part of the tendon; in what structure might this fluid have accumulated?
4 Why was the pain worst during the push-off element of the stance phase?
5 No pain was reported during standing; is this because the tendo calcaneus was relaxed?

YOUR ANSWERS

1 Muscles inserted via tendo calcaneus

2 Reason muscle group called triceps surae

3 Structure in which fluid accumulated

4 Reason for pain being worst during push-off

5 Reason why pain absent on standing

See McMinn pp. 332, 336

Further reading
Apley, A. G. and Solomon, L. *Apley's System of Orthopaedics and Fractures.* 7th edn. Butterworth-Heineman, Oxford, 1993.
Klenerman, L. *The Foot and its Disorders.* 3rd edn. Blackwell Scientific Publications, Oxford, 1991.

CASE 50

Mr George Cardoness, a 49-year-old university chief technician, has been troubled by athlete's foot, particularly in the clefts between the lateral three toes. It became particularly bad when on holiday recently. Then, after returning from a hike in the Scottish hills, Mr Cardoness noticed tender enlarged lymph nodes in his groin. About an hour after returning to his caravan, he started to feel cold and was shivering, even though it was a warm day. Red streaks appeared on the medial side of Mr Cardoness's calf, and he lost his appetite, felt tired and was unable to concentrate on reading or the television.

He called the local doctor who explained that the skin, infected with the fungus responsible for athlete's foot, must have become secondarily infected with bacteria. The doctor prescribed a 10-day course of flucloxacillin and sodium fucidate ointment to control the bacterial infection. Nystatin ointment was then used to treat the athlete's foot.

QUESTIONS

1 What was the cause of the red streaks on the medial side of the calf?
2 Which lymph nodes lie at the groin and how are they arranged?
3 What region of skin drains to the popliteal lymph nodes?
4 To which lymph nodes does lymph from the skin of the lateral toes first drain?
5 Do the lymph nodes at the groin drain sites other than the lower limb?

YOUR ANSWERS

1 Cause of red streaks on calf

2 Name and describe lymph nodes in groin

3 Skin area drained by popliteal lymph nodes

4 Lateral toe skin drained direct by which lymph nodes

5 Other sites drained by groin lymph nodes

See McMinn pp. 67–69, 247–248, 343

Further reading
MacKie, R. M. *Clinical Dermatology: An Illustrated Textbook.* 3rd edn. Oxford Medical Publications, Oxford, 1991.
Williams, P. L., Warwick, R., Dyson, M. and Bannister, L. H. *Gray's Anatomy.* 37th edn. Churchill Livingstone, Edinburgh, 1989.
Yoffey, J. M. Lymphatic system. In: Hamilton, W. J. (ed.) *Textbook of Human Anatomy.* MacMillan Press Ltd., London, 1976, pp. 279–295.

CASE 51

Mrs Christina McGraw, aged 75 and a smoker since 16, was admitted to the vascular surgery unit with a four-week history of pain in the second toe of her right foot. It had gradually got worse over the four weeks prior to admission. The pain was constant, but made worse by lying down in bed; a hot-water bottle made it very painful. Relief was obtained by getting out of bed and placing the toe on something cold, such as the linoleum in the kitchen. Mrs McGraw could hardly walk for the pain.

Further questioning revealed a four-year history of intermittent claudication (pain in the legs on walking) and angina pectoris, relieved by glyceryl trinitrate.

On examination, the second and third toes on the right foot were cyanosed (blue) and very tender. On both lower limbs, only the femoral pulses were palpable and the feet were cold. Buerger's test was positive (feet became pale on elevation).

An arteriogram of the limbs is shown (*Fig. 6.1*). For comparison, an arteriogram from a relatively healthy patient is also shown (*Fig. 6.2*).

QUESTIONS

1 What pulses, other than the femoral, are readily palpable in the healthy lower limb?
2 On the arteriogram of the healthy limb, identify A to D
3 Which vessels are seen on Mrs McGraw's arteriogram?

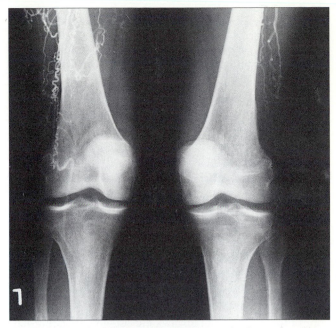

Fig. 6.1 Arteriogram of Mrs McGraw's limbs.

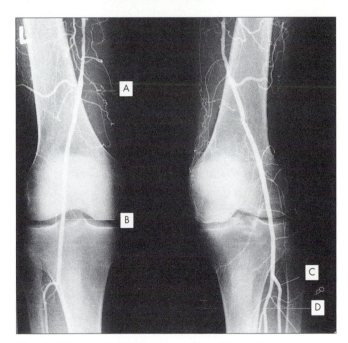

Fig. 6.2 Normal arteriogram of healthy patient.

YOUR ANSWERS

1 Pulses palpable in lower limb, besides femoral pulse

2 Identification of features A to D

3 Vessels on Mrs McGraw's arteriogram

See McMinn pp. 314, 317–318, 328, 331, 332, 335, 339

Further reading
Talley, N. and O'Connor, S. *Clinical Examination.* 2nd edn. Blackwell Scientific Publications, Oxford, 1992.
Yao, J. S. T. and Neiman, H. L. Occlusive arterial disease below the inguinal ligament. In: Neiman, H. L. and Yao, J. S. T. (eds.) *Angiography of Vascular Disease.* Churchill Livingstone, New York, 1985, pp. 109–150.

CASE 52

Mr William Goldie, a 50-year-old unemployed former steel worker, had a four-year history of bilateral osteoarthritic hips. About nine months ago, however, there had been a gradual increase in left-sided hip pain around the greater trochanter, which sometimes radiated distally to the knee. No cause had been identified.

Pain was increased by standing, walking, and lying on the left side, but especially by climbing upstairs. The pain was a constant dull ache, which increased intermittently with any of the previously mentioned activities to a sharp shooting pain. This soon settled within a few minutes to the dull ache again.

All movements were limited by pain on the left side. Resisted isometric testing (pushing against resistance) produced pain on hip abduction and extension. Other isometric tests were negative with regard to pain. However, the power of hip extension was markedly reduced and there was little contraction of the lateral rotators. Some wasting and reduced tone of the left buttock muscles were also noted. During gait, a reduction in stance phase and a positive Trendelenburg's sign could be seen on the left side. On palpation, the area around the greater trochanter was extremely painful, especially towards the posterior aspect.

The diagnosis was of trochanteric bursitis. Treatment was by ultrasound, active strengthening exercises, and pain relief.

QUESTIONS

1 What is a bursa?
2 Where does the trochanteric bursa lie?
3 What is the action of the gluteus maximus muscle on the hip joint?
4 Why should the pain have been especially bad on climbing stairs?
5 What is Trendelenburg's sign? Why do the abductors of the hip of the supporting limb contract during walking?

YOUR ANSWERS

1 Bursa

2 Site of trochanteric bursa

3 Gluteus maximus action on hip joint

4 Reason pain bad climbing stairs

5 Trendelenburg's sign, and reason hip abductors of supporting limb contract during walking

See McMinn pp. 314, 318, 319, 321–323, 340–342

Further reading
Abernethy, P. J. The locomotor system. In: Munro, J. and Edwards, C. (eds.) *Macleod's Clinical Examination.* 8th edn. Churchill Livingstone, Edinburgh, 1990, pp. 255–296.
Pinals, R. S. Traumatic arthritis and other conditions. In: McCarty, D. J. (ed.) *Arthritis and Allied Conditions.* 11th edn. Lea and Febiger, Philadelphia, 1989, pp. 1371–1389.

CASE 53

Mrs Doris Nesbit, a 61-year-old retired school-teacher, kneels a lot while doing her housework. For some weeks, since tripping up on an uneven footpath and grazing her left knee, she had noticed a slight, intermittent, swelling at her left kneecap.

On the day following her fall, a localised swelling had developed in front of the lower part of the patella. The skin in the region became bluish and slightly warm. The swelling became rounded, about 3 cm in diameter, and fluctuant (*Fig. 6.3*).

Over the subsequent few days, the lump got bigger, but was not painful. On the fifth day after the fall, while bending to clean some stairs, Mrs Nesbit saw a small jet of 'whisky-coloured' fluid burst from the swelling, which consequently became softer and smaller.

Over the next few weeks, there was a history of intermittent increases in the size of the swelling, followed by leakages of fluid. Mrs Nesbit used a cotton pad to protect the knee, but was aware of her leg becoming wet when the fluid escaped.

The condition gradually resolved over subsequent months. At no time was the swelling painful, nor did it leak pus.

QUESTION

From your knowledge of anatomy, what was the likely diagnosis?

YOUR ANSWER

Diagnosis

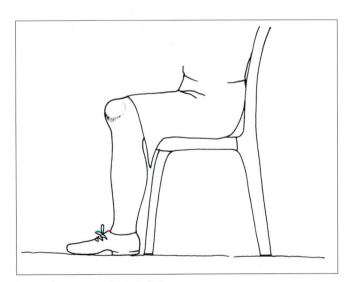

Fig.6.3 Mrs Nesbit's left knee.

Further reading
Apley, A. G. and Solomon, L. *Apley's System of Orthopaedics and Fractures.* 7th edn. Butterworth-Heinemann, Oxford, 1993, p. 463.
Browse, N. L. *An Introduction to the Symptoms and Signs of Surgical Disease.* 2nd edn. Edward Arnold, London, 1992.
Clain, A. *Hamilton Bailey's Demonstrations of Physical Signs in Clinical Surgery.* 17th edn. Wright, Bristol, 1986.

CHAPTER 7

CENTRAL NERVOUS SYSTEM

CASE 54

Mr Duncan McRoberts, an 80-year old retired railwayman, was admitted to hospital having suffered a stroke during the night. He had a history of transient ischaemic attacks over the previous year, in which his speech became slurred and walking was difficult. He was also taking medication for hypertension and had smoked 30 cigarettes per day until three years ago.

On examination, Mr McRoberts was alert and orientated. He was finding it difficult to articulate words, but was able to understand when spoken to, and to name objects. The left side of his mouth was drooping. He could close his eyes, though the action was weaker for his left eye. Both eyes were deviated to the right. The right sides of the visual fields of each eye seemed intact, but the left sides were difficult to assess.

The left upper and lower limbs were paralysed. Muscle tone was increased and reflexes brisk on the affected side. Plantar reflexes were upgoing. Sensation was intact on the right side and absent on the left, but was difficult to assess satisfactorily. The diagnosis was clearly a cerebrovascular accident (CVA).

A computed tomography (CT) scan showed a lesion in the posterior limb of the right internal capsule, with older lesions in the right external capsule and the left lentiform nucleus. The appearances of the lesions were consistent with infarctions (death of tissue from loss of blood supply).

QUESTIONS

1 Had the CVA involved the sensory language (Wernicke's) or motor speech (Broca's) area?
2 Why were the muscles of the upper part of the face still active, when the muscles of facial expression at the left side of the mouth were paralysed?
3 Which clinical signs indicated an upper motor neuron lesion?
4 Did the lesions seen on CT scan explain the clinical findings?

YOUR ANSWERS

1 Involvement of sensory language or motor speech area

2 Reason activity persisted in upper face compared to paralysis at left side of mouth

3 Signs indicating upper motor neuron lesion

4 CT lesions

See McMinn pp. 50–51, 142, 195, 199, 205–207

Further reading
Cull, R. E and Will, R. G. Cerebral vascular disorders. In: Edwards, C. R. W. and Bouchier, I. A. D. (eds.) *Davidson's Principles and Practice of Medicine.* 16th edn. Churchill Livingstone, Edinburgh, 1991, pp. 859–868.
Langhorne, P., Williams, B. O., Gilchrist, W. and Howie, K. Do stroke units save lives? *Lancet* **342**, 395–401, 1993.
Markariam, M. F. Communication: speech and language. In: Dittmar, S.(ed.) *Rehabilitation Nursing: Process and Application.* C. V. Mosby Co., St. Louis, Missouri, 1989, pp. 287–333.
Orgogozo, J. M. Y. and Dyken, M. (eds.) *Advances in stroke prevention.* Sanofi Winthrop Symposium, Second European Stroke Conference, Lausanne, Switzerland (June 26, 1992). Kargel, Basel, 1992.
Toole, J. F. *Cerebrovascular disorders.* 4th edn. Raven Press, New York, 1990.

CASE 55

Mr Robert Millar, a 49-year-old customs and excise officer, was admitted for investigation because of giddiness (vertigo) of acute onset, accompanied by clumsiness of movement (ataxia). Over the six months prior to admission, he reported that he had become progressively more unsteady on his feet and tended to fall to the left. There were also periods of numbness in his left hand. He suffered frequent headaches, which he described as being on both sides and inside the front part of his head. There was no other relevant history. The patient smoked 20 cigarettes a day.

Systematic examination revealed no abnormalities except in the nervous system. The range of abduction of the right eye was reduced, but no double vision was experienced since the patient had amblyopia in that eye (lazy eye). Pupillary reflexes to both light and accommodation were sluggish, but there was no nystagmus. When the tongue was protruded, it deviated slightly to the left. No abnormalities of sensation, power or reflexes were detected, but coordination was globally impaired, most markedly in the left arm, and the gait was ataxic.

No firm diagnosis could be made from the above mixed picture of symptoms and signs, but a computed tomography (CT) scan showed hydrocephalus – accumulation of cerebrospinal fluid (CSF) – secondary to stenosis of the cerebral aqueduct. No cause of the stenosis was apparent.

Mr Millar was treated by the insertion of a ventriculoperitoneal shunt. This is a fine tube used to drain CSF from one of the lateral ventricles, usually the right, to the peritoneal cavity.

QUESTIONS

1 Palsy of which nerve is indicated by the reduction of abduction of the right eye?
2 Where in the brain is the cerebral aqueduct situated?
3 Which structure produces CSF, and where is it situated?
4 Between which meningeal layers does CSF percolate?
5 Where, and by which structures, is CSF resorbed into the bloodstream?
6 How does CSF exit from the fourth ventricle?
7 As the cerebral aqueduct was blocked, which ventricles would probably have been distended?

YOUR ANSWERS

1 Nerve palsy suggested by reduced right-eye abduction

2 Site of cerebral aqueduct

3 Source of CSF and its site

4 Meninges between which CSF percolates

5 Site and structures of CSF resorption into bloodstream

6 CSF exit from fourth ventricle

7 Ventricles distended by cerebral aqueduct blockage

See McMinn pp. 132, 133, 167–168, 196

Further reading
Bradshaw, J. R. *Brain CT: An Introduction.* Wright, Bristol, 1985.
Harrison, M. J. G., Robert, C. M. and Uttley, D. Benign aqueduct stenosis in adults. *J Neurol Neurosurg Psychiat* **37**, 1322–1328, 1974.
McCullough, D. C. Hydrocephalus: treatment. In: Wilkins, R. H. and Rengachary, S. S. (eds.) *Neurosurgery.* McGraw-Hill Book Co., New York, 1985, pp. 2140–2150.
Quest, D. O. Increased intracranial pressure, brain herniation, and their control. In: Wilkins, R. H. and Rengachary, S. S. (eds.) *Neurosurgery.* McGraw-Hill Book Co., New York, 1985, pp. 332–342.

CASE 56

The neurosurgery unit admitted 75-year-old Mrs Margaret Gibson, who had a one-month history of increasing confusion and impaired mobility, and a one-week history of urinary incontinence and mild weakness of the left arm. On the day prior to admission she had become drowsy, and on arrival at hospital she had a mild left hemiparesis and upgoing plantar reflexes. On careful questioning, Mr Gibson recalled that his wife had hit the back of her head about a month previously, while getting up too quickly from under a table.

Twenty-four hours after admission, Mrs Gibson's conscious level, as monitored using the Glasgow coma scale, was falling. She responded to pain only by opening her eyes, flexing her limbs, and with confused words. She had developed papilloedema (swelling of the optic disc) in the right eye.

Computed tomography (CT) scan showed a chronic subdural haematoma (blood clot) extending over much of the lateral side of the right hemisphere. This was causing midline shift and dilatation of the contralateral (opposite side) lateral ventricle. One of the views is diplayed here (*Fig. 7.1*).

Twin burrholes were used to evacuate the clot and a brain catheter provided further drainage. Mrs Gibson made a slow postoperative recovery and was discharged home after a week. The neurological deficits had resolved, although mild confusion persisted.

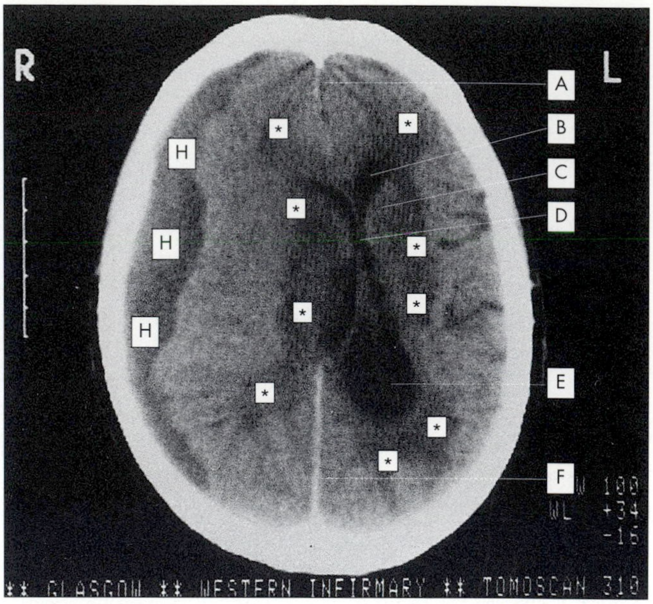

QUESTIONS

1 The subdural space lies between which meningeal layers? What does the subdural space contain in a healthy person?
2 From which vessel(s) had the haemorrhage originated?
3 What does papilloedema indicate and how does it occur?
4 The haematoma has distorted the features of the brain, but can you identify features A to F?

Fig. 7.1 Computed tomography (CT) scan of Mrs Gibson's chronic subdural haematoma: H = haematoma; * = cerebral oedema.

YOUR ANSWERS

1 Meninges containing subdural space, and contents of space in healthy person

2 Origin of haemorrhage

3 Significance and cause of papilloedema

4 Identification of features A to F

See McMinn pp. 134, 166, 198

Further reading
Bradshaw, J. R. *Brain CT: An Introduction.* Wright, Bristol, 1985.
Brihaye, J. Chronic subdural hematoma. In: McLaurin, R. C. (ed.) *Extradural Collections.* Springer-Verlag Wien, New York, 1986, pp. 101–156.
Miller, J. D. Head injury. In: Miller, J. D. (ed.) *Northfield's Surgery of the Central Nervous System.* 2nd edn. Blackwell Scientific Publications, Edinburgh, 1987, pp. 795–870.
Teasdale, G. and Jennett, B. Assessment of coma and impaired consciousness: a practical scale. *Lancet* **ii**, 81–84, 1974.

CASE 57

Mrs Margaret McNeil, a 68-year-old retired school-dinner lady, was admitted to the general medicine ward with weight loss, lack of appetite, anaemia and cardiac failure. The patient's main complaint was pain, but she was rather depressed and seemed unable to describe its site and character. She had been prescribed indomethacin some months before for a sore back and had been taking this in large doses for the pain.

A full blood count indicated that the patient was anaemic and biochemical analysis showed deranged levels of urea, electrolytes, calcium and phosphate. Renal ultrasound and an intravenous pyelogram showed both kidneys to be atrophic. These results led to a diagnosis of renal failure, secondary to excessive dosage of indomethacin. The patient died after two weeks in hospital in which the renal failure progressively worsened.

Throughout her time on the ward, Mrs McNeil complained of pain, but was unable to say exactly where the pain was. Although there was no history of a cerebrovascular accident, the medical staff had wondered if she had 'thalamic syndrome'. This is a condition in which a lesion of the pain pathways disturbs the neuronal circuitry so that the patient experiences pain without a peripheral cause. As a neurological examination had shown no sensory deficits, a lesion in the thalamus itself was being considered. The patient died before this could be adequately investigated and the relatives did not want a post-mortem performed.

QUESTIONS

1 Where is the thalamus situated and of what is it composed?
2 Is pain the only sensory modality to involve the thalamus?
3 Which nucleus of the thalamus is principally involved as a relay nucleus on the general sensory pathways?
4 To which region of the cortex does this nucleus primarily project?
5 From which part of the circle of Willis does the thalamus receive its blood supply? What general route do these blood vessels take?

YOUR ANSWERS

1 Site and composition of thalamus

2 Sensory modalities involving thalamus

3 Thalamic nucleus relaying sensory signals

4 Cortical region receiving sensory relay

5 Section of circle of Willis supplying blood to thalamus, and circulatory route

See McMinn pp. 193, 196–199, 207–211

Further reading
Andersson, S., Bond, M., Mehta, M. and Swerdlow, M. *Chronic Non-Cancer Pain*. MTP Press Ltd., Lancaster, 1987.
Boivie, J. Central pain. In: Wall, P. D. and Melzack, R. (eds.) *Textbook of Pain*. 3rd edn. Churchill Livingstone, Edinburgh, 1994, pp. 871–902.
Diamond, A. W. and Coniam, S. W. *The Management of Chronic Pain*. Oxford University Press, Oxford, 1991.
Fordham, M. and Dunn, V. *Alongside the Person in Pain: Holistic Care and Nursing Practice*. Baillière Tindall, London, 1994.
Maciewicz, R. and Fields, H. L. (1986) Pain pathways. In: Asbury, A. K., McKhann, G. M. and McDonald, W. I. (eds.) *Diseases of the Nervous System: Clinical Neurobiology*. Vol. II. W. B. Saunders Co., Philadelphia, 1986, pp. 930–940.

CASE 58

A 44-year-old man with a long history of bronchiectasis (permanent dilatation of the bronchi) and chest infections died of bronchopneumonia and respiratory failure. The subject also had a spastic monoparesis of his left upper limb and epilepsy dating from meningitis at age seven.

The post-mortem examination of the brain was remarkable. The brain was small (855 g). There was a large defect in the lateral aspect of the frontal and parietal lobes of the right cerebral hemisphere, within the distribution of the middle cerebral artery. The defect measured 10 cm anteroposteriorly, 6 cm from above down, and on coronal sections was up to 5.5 cm from side to side. The lesion was wedge-shaped, with its base at the cortex, and its apex extending towards the lateral ventricle.

In the central part of the affected area, there was total loss of brain tissue, so that the ependyma of the enlarged body of the right lateral ventricle was in contact with the meninges. There were no features in the meninges to suggest that the defect had been caused by an abscess. In this region of the brain, there was loss of cortex and white matter and of parts of the corpus striatum and thalamus.

The parasagittal cortex was intact, as were the anterior part of the frontal lobe, the posterior part of the parietal lobe and the occipital lobe; the temporal lobes, however, were asymmetrical with the right lobe being smallest.

The main arteries at the base of the brain were patent and of normal configuration. The cerebellum and cranial nerves were healthy. However, the right side of the brainstem had atrophied because the corticospinal tract fibres arising from the affected motor cortex had degenerated.

See McMinn pp. 40, 193–195, 196–199, 205–207

Further reading

Fitzgerald, M. J. T. *Neuroanatomy: Basic and Clinical.* 2nd edn. Baillière Tindall, London, 1992.

QUESTIONS

1 Which region(s) of the brain are supplied by the middle cerebral artery?
2 Is the damage consistent with blockage of the entire middle cerebral artery, or only with some of its branches?
3 The patient had motor loss in the left upper limb; can you explain why the innervation of the head and lower limb seemed intact clinically?
4 Given that the internal capsule seemed intact, how might the degenerations in the corpus striatum and thalamus be explained?
5 What was the likely cause of the infarction?
6 Where do corticospinal fibres from the right motor cortex lie in sections, at the following levels, in a non-pathological brainstem and spinal cord:
• Midbrain.
• Pons.
• Upper medulla.
• Cervical spinal cord?
7 At what level in the brainstem does the pyramidal (motor) decussation occur?

YOUR ANSWERS

1 Regions supplied by middle cerebral artery

2 Extent of blockage of middle cerebral artery

3 Reason for clinically intact innervation of head and lower limb

4 Reason for degenerations in corpus striatum and thalamus

5 Cause of infarction

6 Levels at which corticospinal fibres lie in sections in healthy brainstem and spinal cord

7 Level of pyramidal decussation

CHAPTER 8

VERTEBRAL COLUMN

CASE 59

Mr Iain Docherty, a 38-year-old shipyard engineer, attended the physiotherapy department with a one-year history of low back pain and intermittent right leg pain, radiating from the posterolateral aspect of the knee to the little toe. Bladder and bowel function were normal. His family doctor had diagnosed a lumbar-disc problem.

Pain was increased by sitting and driving and could be reduced by lying flat. Straight-leg raise was 90° on the left but only 35° on the right. There was no sensory deficit, power was full on resisted testing (pushing against resistance), plantar reflexes were down-going, and knee-jerk reflexes were normal, but the right ankle-jerk reflex was absent.

Observation of the spine showed a scoliosis, concave to the right. Flexion of the lumbar spine was extremely limited and increased leg pain. Extension reduced leg pain but increased central back pain.

Treatment by daily lumbar traction was commenced. Both objective and subjective signs were monitored over two weeks and, though straight-leg raising increased slightly to 40°, there was no change in the symptoms. Consequently, Mr Docherty was referred through his family doctor to an orthopaedic surgeon. Magnetic resonance imaging (MRI) demonstrated L5/S1 posterolateral disc protrusion. The prolapsed disc was excised.

QUESTIONS

1 The cause of Mr Docherty's symptoms was a prolapse of the intervertebral disc between L5 and S1. What is the structure of an intervertebral disc and what happens if it prolapses?
2 Damage to which spinal nerve(s) is suggested by the absence of the right ankle-jerk reflex?
3 Which dermatome would most probably include the posterior aspect of the calf and the lateral side of the foot, including the little toe?
4 Which spinal nerve(s) emerges from the intervertebral foramina between L5 and S1?
5 How does S1 emerge from the vertebral canal?
6 Where do ventral and dorsal roots unite to form the spinal nerves?
7 Why might S1 rather than L5 have been affected in Mr Docherty?

YOUR ANSWERS

1 Structure of normal and prolapsed intervertebral disc

2 Spinal nerve(s) damaged

3 Dermatome involving posterior aspect of calf and lateral side of foot, including little toe

4 Spinal nerve(s) emerging from L5/S1 intervertebral foramina

5 Emergence of S1 nerve from vertebral canal

6 Site where ventral and dorsal roots join to form spinal nerves

7 Reason S1, not L5, might have been affected

See McMinn pp. 27, 40–42, 45, 199, 202–203

Further reading
Apley, A. G. and Solomon, L. *Apley's System of Orthopaedics and Fractures.* 7th edn. Butterworth-Heinemann, Oxford, 1993.
Cohen, M. S., Wall, E. J., Olmarker, K., Rydevik, B. C. and Garfin, S. R. Anatomy of the spinal nerve roots in the lumbar and lower thoracic spine. In: Rothman, R. H. and Simeone, F. A. (eds.) *The Spine.* W. B. Saunders Co., Philadelphia, 1992, pp. 99–106.
Wisneski, R. J., Garfin, S. R. and Rothman, R. H. Lumbar disc disease. In: Rothman, R. H. and Simeone, F. A. (eds.) *The Spine.* W. B. Saunders Co., Philadelphia, 1992, pp. 671–746.

CASE 60

Mr Alan Porter, a 53-year-old lorry driver, had been troubled with intermittent headaches in the occipital region during the last six months. They tended to begin when he was driving, or when relaxing in the evening. The pain would increase in intensity over the period of approximately an hour until Mr Porter stopped driving and took painkillers. Lying down also brought relief. He had no other symptoms. Radiographs had demonstrated slight wear and tear of the small synovial joints between the cervical vertebral bodies, but nothing else. The radiologist questioned the significance of this finding in relation to the patient's pain.

When Mr Porter was sitting, it was noticed that he sat with complete loss of the lumbar lordosis and with his cervical spine protracted, i.e. extended in its upper part and flexed in its lower part (*Fig. 8.1*). When standing, his posture was much improved. The movements of flexion and extension were full and pain-free. Similarly, rotations to the right and left were pain-free; however, they were equally limited at approximately half the range of movement. All other objective assessments, including tests for vertebral artery insufficiency, were negative.

Once she had completed the assessment, the physiotherapist was able to demonstrate that it was Mr Porter's poor sitting posture which induced the headaches. Mr Porter was treated by postural correction to maintain lumbar lordosis and straighten the cervical spine. The patient practised by sitting with a towel behind his lumbar spine and tucking in his chin. In this corrected posture, the range of rotation of the neck was markedly increased. As Mr Porter became more aware of his posture, he was able to prevent the occurrence of the headache or facilitate its removal by the exercises.

QUESTIONS

1 What classes of joints are there between typical cervical vertebrae?
2 What are the normal curvatures of the vertebral column? What is responsible for these curvatures?
3 What movements occur between typical cervical vertebrae? What movements occur at the atlanto-occipital and the atlanto-axial joints?
4 Where does the vertebral artery lie in relation to the cervical vertebrae?
5 How might the function of the vertebral artery be assessed?
6 Why was the headache felt in the occipital region?

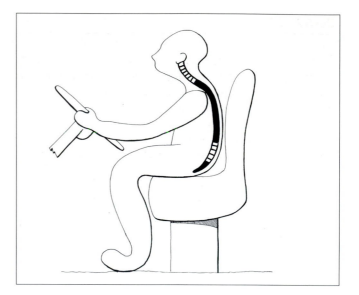

Fig. 8.1 Diagram of Mr Porter's posture showing the protraction of the neck and loss of lumbar lordosis.

YOUR ANSWERS

1 Classes of joints between cervical vertebrae

2 Normal curvatures of vertebral column and origin

3 Movements between cervical vertebrae, and at atlanto-occipital and atlanto-axial joints

4 Vertebral artery in relation to cervical vertebrae

5 Assessment of vertebral artery

6 Reason headache in occipital area

See McMinn pp. 25–26, 31, 174, 178, 212

Further reading
McKenzie, R. *The Cervical and Thoracic Spine: Mechanical Diagnosis and Therapy.* Spinal Publications Ltd., New Zealand, 1990.
Murali, R. Neurosurgical considerations in headaches. In: Jacobsen, A. L. and Donlon, W. C. (eds.) *Headache and Facial Pain: Diagnosis and Management.* Raven Press, New York, 1990, pp. 245–271.

CASE 61

Eighteen months ago, Ms Audrey Garrioch, a 39-year-old company director, was putting dishes into her dishwasher, when she felt a sharp pain in her lumbar spine and a sensation 'like a click'. The next morning, the pain was not too bad, but she went to the chemist for painkillers. While she was in the shop, severe pain came on in her lower back, and she could hardly walk to the car to drive home. The pain radiated to the left thigh, and down the leg to the foot towards the big toe. It was relieved by lying still.

The family doctor prescribed analgesics and, suspecting a prolapsed intervertebral disc, advised complete bedrest for three weeks apart from going to the bathroom. As the days passed, however, Ms Garrioch's symptoms were not resolving and straight-leg raising was not improving. She was therefore admitted to the orthopaedic ward.

No abnormalities were seen on radiology of the lumbar spine.

The history of 'feeling a click' suggested a diagnosis of a foreign body, such as a fragment of cartilage, in a zygapophyseal (facet) joint. Ultrasound and rotational movements were begun to try and separate the facet joint surfaces enough to free the foreign object. Ultrasound physiotherapy brought rapid and marked improvement, thus confirming the diagnosis.

The site of the pain, and its radiation, suggested that the affected joint might be the left zygapophyseal joint between L4 and L5. Ms Garrioch was off work for 10 weeks, during which she had daily and, later, twice weekly physiotherapy.

Since then, Ms Garrioch has had three similar episodes and she is starting to recognise a pattern. The episodes have tended to start with a stiff back in the morning, followed by pains in the back 'like labour pains' and a feeling of her lower back being stretched by muscle spasms. The symptoms have usually settled after 2–3 weeks' treatment with analgesics and ultrasound.

The third time it happened, Ms Garrioch recalled being at a meeting the previous evening and leaning over papers spread out on a coffee-table. Clearly, the episodes are very painful and inconvenient as Ms Garrioch has a busy work schedule, and they have meant taking time off unexpectedly and for relatively long periods.

QUESTIONS

1 Are zygapophyseal (facet) joints found throughout the vertebral column, or are they restricted to the lumbar spine?
2 Between which parts of adjacent vertebrae do zygapophyseal joints lie?
3 To which class of joints do zygapophyseal joints belong?
4 What movements take place in the lumbar spine?
5 What features of the radiation of the pain suggest that the left-sided joint between L4 and L5 was the source of the pain?

YOUR ANSWERS

1 Sites of zygapophyseal joints

2 Parts of adjacent vertebrae enclosing zygapophyseal joint

3 Classification of zygapophyseal joint

4 Movements of lumbar spine

5 Pain source and radiation of pain

See McMinn pp. 21–25, 27, 40–42, 44, 45–46

Further reading
Eisenstein, S. M. and Parry, C. R. The lumbar facet arthrosis syndrome. *J Bone Joint Surg* **69B**, 3–7, 1987.
Latif, N. A. and Shaw Dunn, J. Innervation of the lumbar zygapophyseal joints. *Clin Anat* **6**, 58–59, 1993.
Mooney, V. and Robertson, J. Facet syndrome. *Clin Orthopaed* **115**, 149–156, 1976.

A n s w e r s

CASE 1

1. The external acoustic meatus is appreciably shorter in a child than in an adult and the tympanic membrane is therefore closer to the surface.
2. Respiratory epithelium.
3. The auditory tube communicates with the anterior wall of the middle-ear cavity.
4. Lateral side of nasopharynx, surrounded by tubal elevation.
5. Air in the middle-ear cavity is gradually absorbed. The auditory tube allows new air to enter, thus equalising pressure on each side of the tympanic membrane.
6. Small synthetic tube placed in short incision in tympanic membrane which allows air to enter middle-ear cavity from external acoustic meatus.
7. Pharyngeal tonsils (collections of lymphoid tissue). The adenoids had enlarged as a result of the upper respiratory tract infections and were responsible for blocking the opening of the auditory tube.

CASE 2

1. Herpes zoster (shingles). The condition is caused by the same virus as chickenpox (varicella zoster). Following chickenpox, the virus remains dormant in the dorsal root ganglia of spinal nerves and the sensory ganglia of cranial nerves. At some later time, the virus may be reactivated (by having had influenza in Miss Allen's case) and travel down the sensory nerve to produce the rash in the dermatome of the nerve.
2. Ophthalmic nerve. Herpes zoster usually only affects the skin supplied by one division of the trigeminal nerve. The reason for this is not understood. The ophthalmic nerve is the division most commonly affected.
3. Supraorbital nerve, principal branch of the frontal nerve.
4. External nasal nerves, derived from the nasociliary nerve.
5. Corneal scarring.
6. Long ciliary nerves, branches of the nasociliary nerve.

CASE 3

1. The sensory fibres of the vestibulocochlear nerve convey information about sound and the position and movement of the head from the inner ear.
2. Schwann cells are the supporting glial cells of peripheral nerves.
3. Facial nerve, vestibulocochlear nerve and labyrinthine vessels.

4. Facial nerve. In the posterior cranial fossa, the facial nerve lies alongside the vestibulocochlear nerve.
5. Both attach to the brainstem at the cerebellopontine angle.
6. The facial palsy resulted in paralysis of the ipsilateral muscles of facial expression, including the orbicularis oculi muscle, and therefore blinking was impaired. The tarsorrhaphy was necessary to protect the cornea.
7. Paralysis of the muscles around the left side of the mouth, including the orbicularis oris and buccinator muscles, made it difficult to hold the false teeth in position.

CASE 4

1. The sympathetic trunk, on each side, extends from upper cervical regions to the ganglion impar, anterior to the coccyx.
2. Middle cervical and cervicothoracic (stellate) ganglia. The former is small and lies close to the inferior thyroid artery. The latter is relatively large and related to the neck of the first rib; some individuals show separate lower cervical and first thoracic ganglia.
3. Healthy sympathetic postganglionic neurons produce noradrenaline as a neurotransmitter. Neuroblastoma cells produce similar substances.
4. Superior tarsal muscle. It consists of smooth muscle and assists the levator palpebrae superioris muscle in supporting the upper eyelid.
5. Other signs of sympathetic denervation:
- Constricted pupil on the affected side (paralysis of the dilator pupillae muscle and unopposed action of the sphincter pupillae muscle).
- Ipsilateral loss of sweating.
  Collectively, unilateral ptosis, pupillary constriction and loss of sweating are known as Horner's syndrome.

CASE 5

1. The medial pterygoid muscle is a thick quadrilateral muscle, with a large deep head from the medial side of the lateral pterygoid plate and a small superficial head from the maxillary tuberosity. Its fibres run downwards, backwards and laterally, to reach the medial side of the lower part of the ramus and the angle of the mandible.
2. The mandibular foramen lies in the middle of the medial side of the ramus.
3. The inferior alveolar nerve is a branch of the posterior division of the mandibular nerve in the infratemporal fossa.

4. The inferior alveolar nerve lies between the lateral and medial pterygoid muscles, and approaches the mandibular foramen, beneath the lower border of the lateral pterygoid muscle. The medial pterygoid muscle is below and medial to the nerve close to the foramen.

CASE 6

1. The pterygomandibular raphe is the fibrous band which runs from the pterygoid hamulus to the mandible, medial to the third lower molar tooth.
2. The lower lip receives its sensory supply from the mental nerve, a branch of the inferior alveolar nerve. The lingual nerve, which gives general sensation to the anterior two-thirds of the tongue, lies close to the inferior alveolar nerve in the pterygomandibular space and is therefore also anaesthetised.
3. The patient could be asked to smile, to pucker the lips as in whistling, and to blow the cheeks up and hold the air in them while the examiner taps each cheek with a finger.
4. The palpebral fibres of the orbicularis oculi muscle.
5. It was likely that the blink reflex had been lost. The cornea was at risk of drying and from being struck by any object approaching the eye.
6. The needle had been placed too far posteriorly, and as a result the anaesthetic was injected into the parotid gland. The capsule of the gland prevented the anaesthetic reaching the inferior alveolar and lingual nerves. The anaesthetic solution, however, affected the facial nerve which passes through the gland.

CASE 7

1. No, the other upper teeth are innervated by the anterior and middle superior alveolar nerves.
2. The posterior superior alveolar nerve enters the posterior (infratemporal) aspect of the maxilla.
3. The pterygoid plexus is a rich plexus of veins in the infratemporal fossa and lies in the fat adjacent to the pterygoid muscles.
4. The pterygoid plexus communicates with the cavernous sinus via the sphenoidal emissary foramen or the foramen ovale. Infections may spread by this route to the cranial cavity.

CASE 8

1. Anterior cranial fossa.
2. Arachnoid mater, dura mater, cribriform plate of the ethmoid bone, olfactory epithelium.
3. A full neurological examination would be mandatory, with the olfactory nerves at particular risk. They are tested by assessing the patient's ability to detect mild odours.
4. Tears reach the nasal cavity via the superior and inferior lacrimal canaliculi, the nasolacrimal sac, and the nasolacrimal duct, which opens into the inferior meatus of the nasal cavity.

CASE 9

1. Damage to the sixth (abducens) nerve produced paralysis of the lateral rectus muscle.

2. The vestibular apparatus and proprioceptors in the neck muscles.
3. The medial longitudinal bundle.
4. The raised intracranial pressure may push the medulla downwards, thus stretching the thin abducens nerve as it passes from the pontomedullary junction to pierce the dura mater, over the clivus of the posterior cranial fossa.
5. The abducens nerve is deeply placed and unlikely to have been directly damaged by the forceps. After piercing the dura, it crosses the superior border of the petrous temporal bone, and runs with the internal carotid artery through the cavernous sinus, before entering the orbit via the superior orbital fissure. The external aspect of the greater wing of the sphenoid bone, which forms the lateral orbital wall, is deep to the temporalis muscle.

CASE 10

1. An oro-antral fistula has formed between the socket of the third molar and the maxillary sinus.
2. The roots of the teeth from the first premolar to the third molar, and occasionally the canine, may be related to the maxillary sinus.
3. In some individuals, as much as half the length of the roots may lie in the sinus wall. In such a position, the roots are separated from the sinus mucosa by only a paper-thin layer of bone, which may be missing in places.
4. The ostium of the maxillary sinus is high on its medial wall and opens into the middle meatus of the nasal cavity at the lower end of the hiatus semilunaris. From here the fluid would trickle towards the nostril.

CASE 11

1. The trauma to the face had damaged the infraorbital nerve as it emerged from its foramen.
2. The bones comprising the facial skeleton are:
• Frontal bone.
• Nasal bones.
• Zygomatic bones.
• Maxillae.
• Mandible.

CASE 12

1. Thyroglossal duct.
2. The mass was probably tethered to the hyoid bone.
3. Fluctuation is the ability to change shape if compressed. The cyst was non-fluctuant because the fluid was under pressure.
4. The cyst indicated a minor defect of thyroid development. The surgeon was checking that thyroid tissue had not persisted in the tongue (lingual thyroid).
5. Foramen caecum. Pyramidal lobe and levator glandulae thyroideae muscle.

CASE 13

1. The thyroid gland has right and left lobes connected by an isthmus. The lobes extend from the oblique line on the thyroid

cartilage to about the fourth or fifth tracheal ring. The isthmus lies anterior to the second, third and fourth tracheal rings.
2. The thyroid gland is connected to the larynx and trachea by the pretracheal fascia.
3. Thyroid hormone comprises tetra-iodothyronine/ thyroxine (T4) and tri-iodothyronine (T3). Production of both hormones requires iodine uptake.
4. No; while they all arise from the primitive pharynx, they are initially not intimately related. The thyroid arises from the floor of the primitive pharynx, whereas the superior and inferior parathyroid glands arise from the fourth and third pouches, respectively. The parathyroids attach to the thyroid gland during their descent in the neck.
5. Deep cervical nodes.
6. The external laryngeal nerve and the recurrent laryngeal nerve are related to the superior and inferior thyroid arteries, respectively; the nerves may be damaged during ligation and division of these vessels.

CASE 14

1. Left unilateral cleft lip and palate.
2. The maxillary and medial nasal processes have failed to fuse.
3. In first year, sucking; from second year onwards, speaking.

CASE 15

1. The cornea is the main site of refraction; the lens provides adjustments of the focus.
2. The lens capsule is a thick, transparent, basement membrane which surrounds the lens. It is suspended from the ciliary body by the zonule fibres.
3. Access of aqueous humour to the canal of Schlemm (sinus venosus sclerae) at the iridocorneal angle is impaired. The poor drainage leads to a rise in intra-ocular pressure.
4. The trabecular meshwork is a lattice of fibrous trabeculae through which aqueous humour percolates to reach the canal of Schlemm.
5. Pilocarpine stimulates acetylcholine receptors to produce contraction of the sphincter pupillae muscle and to open up the trabecular meshwork for aqueous drainage. The ciliary muscle is also likely to be stimulated and this may interfere with accommodation.

CASE 16

1. The supraspinatus and infraspinatus muscles attach to the upper and middle facets, respectively, of the greater tubercle. The subscapularis muscle attaches to the lesser tubercle. The teres minor muscle attaches to the lower facet of the greater tubercle.
2. The supraspinatus muscle initiates abduction and is active in the early stages of the movement; this corresponds with the pain on resisted abduction. The painful arc is probably unrelated to the action of the supraspinatus muscle and represents compressive forces on the tender tendon.
3. The first 30° of elevation occurs at the glenohumeral joint only. Thereafter, about two-thirds of the movement occurs at the shoulder joint and about one-third by scapular rotation.

4. The supraspinatus muscle is an abductor of the shoulder joint, the subscapularis muscle is a medial rotator of the humerus, while the infraspinatus and teres minor muscles are lateral rotators. Together, they form the rotator cuff which confers stability on the joint.
5. The suprascapular nerve which supplies the supraspinatus muscle contains C5 fibres. The pain was referred to the dermatome of C5.

CASE 17

1. Serratus anterior muscle. This muscle passes backwards from the upper eight ribs to insert along the deep aspect of the medial border of the scapula. When this muscle is paralysed, the medial border of the scapula is no longer held against the chest wall and winging occurs.
2. The lower fibres of the serratus anterior muscle, together with the upper fibres of the trapezius muscle, produce lateral rotation of the scapula. This extends the range by which the limb can be raised.
3. Pushing against the wall tests the power of protraction – a movement in which the serratus anterior muscle is assisted by the pectoralis minor muscle.

CASE 18

1. The radial styloid process is the bony projection at the lateral side of the distal end of the radius.
2. The bones that are part of the wrist (radiocarpal) joint include:
 • Distal end of the radius.
 • Scaphoid bone.
 • Lunate bone.
 • Triquetral bone.
 The head of the ulna is excluded from the wrist by the articular disc.
3. Ellipsoid joint. Flexion, extension, abduction, adduction and circumduction.
4. The radiocarpal and intercarpal joints work as a functional unit producing about 85° of flexion and about 85° of extension.
5. The superior and inferior radio-ulnar joints.
6. Saddle joint. Flexion, extension, abduction, adduction and opposition.

CASE 19

1. As a result of the atrial fibrillation, an embolus (blood clot) had formed in the heart and passed into, and blocked, the right brachial artery.
2. The brachial artery is palpated at the cubital fossa, medial to the biceps tendon. The radial artery is felt anterior to the distal end of the radius, lateral to the flexor carpi radialis muscle.
3. It is unlikely that blood would have been flowing in the patient's ulnar artery, as the radial and ulnar arteries are terminal branches of the brachial artery. The ulnar artery is deeply placed and difficult to palpate.
4. The principal blood supply of the upper limb is from the subclavian artery which becomes the axillary artery beyond the first rib. At the lower border of the teres major muscle, the artery

becomes the brachial artery. In the cubital fossa, the brachial artery divides into the radial and ulnar arteries, which continue down the forearm and anastomose in the superficial and deep palmar arches in the palm.

CASE 20

Probably the C8 root of the plexus. Several features suggest this diagnosis:
- History of severe upward traction (pull) on upper limb.
- Pain and tenderness in neck.
- Distribution of paraesthesia (pins and needles) in C8 dermatome.
- Weakness of muscles of forearm not conforming to the individual distribution of the median, ulnar or radial nerves.
- Pain reproduced on minor tension of the median, ulnar and radial nerves, all of which contain C8 fibres.

COMMENT

There was no Horner's syndrome. Damage to the sympathetic nerve supply to the head would have suggested damage to the T1 spinal nerve proximal to the white rami communicantes.

CASE 21

The acromioclavicular joint had subluxed (loosened).

CASE 22

1. The annular ligament is attached at each side of the radial notch on the ulna and surrounds the radial head.
2. The annular ligament contributes to the superior radio-ulnar joint where, in conjunction with the inferior radio-ulnar joint, pronation and supination occur.
3. Olecranon process, lateral and medial epicondyles of the humerus, head of radius.
4. The annular ligament lies between the articular surfaces of the capitulum of the humerus and the head of the radius.

CASE 23

1. The palmar aponeurosis.
2. Dense connective tissue composed of fibroblasts and collagen fibres.
3. The distal part of the palmar aponeurosis gives four slips which blend with the ligaments of the metacarpophalangeal joints and with the fibrous flexor sheaths of the fingers.

CASE 24

1. With the thumb in extension, the anatomical snuff-box is the triangular depression on the dorsum of the hand between the tendons of the extensor pollicis longus muscle medially and the extensor pollicis brevis muscle laterally.

2. The scaphoid bone is palpable in the anatomical snuff-box.
3. The bony features A to E are:
- A = Trapezium.
- B = Scaphoid bone.
- C = Hamate bone.
- D = Triquetral bone.
- E = Ulnar styloid process.
4. The radiograph shows a fracture of the scaphoid. The blood supply of the scaphoid tends to enter the more distal part of the bone. Thus, a scaphoid fracture carries the risk of avascular necrosis of the proximal fragment.

CASE 25

1. In the fetus, the ductus arteriosus is a vascular shunt between the pulmonary artery and the arch of the aorta distal to the origin of the left subclavian artery.
2. At birth, the ductus arteriosus closes under the influence of prostaglandins and forms the fibrous cord, the ligamentum arteriosum.
3. The blood in the ductus arteriosus flows from the pulmonary artery to the aorta.
4. The foramen ovale.

CASE 26

1. In a healthy individual, the pleural cavity is a potential space between the visceral and parietal layers of the pleura. It contains only a thin film of tissue fluid.
2. Air must have entered the pleural cavity through a small rupture on the surface of the lung. The presence of the air caused the hyper-resonance on percussion, and the withdrawal of the lung away from the chest wall led to the loss of vocal fremitus and resonance.
3. Skin and superficial fascia, serratus anterior muscle, external, internal, and innermost intercostal muscles, and parietal pleura.
4. The dome of the diaphragm is usually below this level. The risk of piercing the diaphragm and liver was thereby reduced.
5. The parietal pleura of the chest wall is innervated by the intercostal nerves, while the pleura covering the lungs is supplied by visceral nerves. Pleuritic pain is sharp and varies with respiration; it arises from the parietal pleura.
6. The reason for the swelling and difficulty in swallowing is uncertain, but they were probably caused by enlargement of the tracheobronchial lymph nodes, in response to the associated respiratory tract infection.

CASE 27

1. On the angiogram, the following structures are identified:
- A = Main stem of left coronary artery.
- B = Anterior interventricular (left anterior descending) branch of left coronary artery.
- C = Circumflex branch of left coronary artery.
2. The popliteal pulse is most readily felt by flexing the knee to about 120° and pressing the fingertips into the popliteal fossa; the artery is deeply placed so the pulse is diffuse in nature. The posterior tibial pulse is palpable about 1 cm below and posterior

to the medial malleolus. The dorsalis pedis pulse is felt over the intermediate cuneiform on the dorsum of the foot.

CASE 28

1. There are two lymphatic plexuses in the lung: a superficial one below the pleura and a deep plexus adjacent to the bronchioles and bronchi. Both plexuses drain to the bronchopulmonary lymph nodes at the hilum of the lung.
2. The deposits of tumour had spread beyond the first regional nodes to adjacent groups of lymph nodes.
3. The carbohydrate component of mucus is stained by the periodic acid-Schiff technique (magenta) or by the mucicarmine technique (red).
4. Goblet cells are scattered throughout the pseudostratified columnar epithelium which lines the bronchi. In addition, the submucosal glands contain mucous acini with serous demilunes.
5. The lower part of the superior vena cava lies anterior to the lung root.

CASE 29

1. The right lung has upper, middle and lower lobes, separated by the oblique and horizontal fissures. The left lobe has upper and lower lobes separated by the oblique fissure; the lingula is part of the upper lobe.
2. The oblique fissure lies close to the line of the sixth rib. The left oblique fissure is more vertical than that on the right.
3. The most obvious accessory muscle would be the pectoralis major muscle; the patient might well grasp the bed to fix the upper limbs. Contraction of the sternocleidomastoid muscle would also be readily observed.
4. No. The costodiaphragmatic recess would reach to the 12th rib. The lung bases would be higher, about two ribs above the costal margin.
5. A bronchopulmonary segment is an independent unit of lung parenchyma, supplied with air by a segmental bronchus, a branch of a lobar bronchus.
6. The number of bronchopulmonary segments in each lobe:
   - Right upper lobe, 3.
   - Right middle lobe, 2.
   - Right lower lobe, 5.
   - Left upper lobe, 5 (includes two lingular segments).
   - Left lower lobe, 5.
7. Much of quiet inspiration is brought about by contraction of the diaphragm. Quiet expiration is brought about by elastic recoil; the diaphragm is relaxed.
8. In quiet inspiration the lower ribs move upwards and outwards (bucket-handle movement).

CASE 30

1. Metastatic tumour in the axillary lymph nodes had blocked the passage of lymph through the nodes.
2. Pectoralis major muscle.
3. The lump was tethered to the muscle and became fixed when it contracted.
4. The superolateral part of the breast extends towards the axilla as the 'axillary tail'.

5. The breast ducts are lined by columnar epithelium. In the larger ducts, this may consist of a double layer of cells. A layer of myoepithelial cells surrounds the columnar cells and rests on the basement membrane. The epithelium is surrounded by connective tissue rich in elastic fibres.
6. The medial parts of the breast drain to the internal thoracic lymph nodes, alongside the internal thoracic blood vessels. The nodes lie to the side of the deep aspect of the sternum and are inaccessible to surgery.
7. Tamoxifen is an anti-oestrogenic drug which binds to the oestrogen receptors and prevents the stimulating action of oestrogen on nucleic-acid synthesis.

CASE 31

1. As the patient was resting in bed, fluid had readily collected in the tissues over the sacrum.
2. The inferior border of the liver usually lies close to the right costal margin; in some healthy individuals, it may be just palpable. Part of the border lies inferior to the infrasternal angle.
3. In a healthy individual, the spleen is never palpable. It lies in the posterior part of the left hypochondriac region, and is separated by the diaphragm and the lower parts of the left lung and pleura from the 9th, 10th and 11th ribs. A normal spleen lies behind the mid-axillary line.
4. The notch lies towards the anterior end of the superior margin.
5. Although the patient's haemorrhoids – distension of the portasystemic anastomoses between the superior and inferior rectal veins within the anal columns – may indicate portal hypertension, haemorrhoids are very common in the general population and may have been an incidental finding.

CASE 32

The features A to G are:
- A = T-tube.
- B = Intrahepatic tributaries of right hepatic duct.
- C = Common hepatic duct.
- D = Stump of cystic duct.
- E = Bile duct.
- F = Contrast medium in descending part of duodenum.
- G = Gallstone in bile duct.

CASE 33

1. The cirrhosis disrupted the structure of the liver and impeded the blood flow through the organ, resulting in portal hypertension. The increased blood pressure in the portal vein and its tributaries caused distension of the portasystemic anastomosis, which connects the left gastric vein and the azygos system of veins of the mediastinum. This anastomosis lies in the lower oesophagus and, when distended, produces oesophageal varices.
2. Hepatocytes excrete bilirubin, a product of red blood corpuscle breakdown, and metabolise other waste products. They also inactivate oestrogen. Loss of hepatocytes had impaired these processes and led to jaundice, confusion and gynaecomastia, respectively.
   The oedema was a result of reduced resorption of tissue fluid into the venous ends of capillaries. Reduced tissue fluid

resorption was itself caused by reduced levels of plasma proteins, which are normally produced by the liver. The poor blood clotting also reflected reduced production of plasma proteins, some of which are clotting factors.

In the abdomen, the decreased resorption of tissue fluid, associated with reduced plasma proteins, was compounded by high venous pressure in the portal vein tributaries. This further decreased absorption of fluid in the abdomen, leading to the condition known as ascites.

3. The splenic vein is a tributary of the portal vein. The portal vein and its tributaries have no valves so that portal hypertension produces venous engorgement of the spleen.
4. The lower oesophagus is the most important site clinically. Portasystemic anastomoses also occur:
- At the bare area of the liver.
- Around the umbilicus.
- Behind the ascending and descending colons.
- In the anal canal.

CASE 34

1. Autosomal dominant inheritance.
2. The enlarged polycystic kidneys.
3. Uric acid; its excretion is impaired by renal failure.
4. The nephrons develop from the metanephric mesoderm, while the collecting ducts develop from the ureteric bud.
5. Shunts are often fashioned just above the wrist, by anastomosing the radial artery to the cephalic vein.

CASE 35

1. An inguinal hernia is an abnormal protrusion of the abdominal contents into the inguinal canal.
2. Crying involves forced expiration, which requires a raised intra-abdominal pressure to elevate the diaphragm. The rise in intra-abdominal pressure caused the hernia to appear.
3. In fetal life, the testis descends from its original position in the abdomen and passes through the inguinal canal. Just before birth, it enters the scrotum. The processus vaginalis, a sleeve-like extension of the peritoneum, accompanies the testis into the scrotum. The part of the processus around the testis persists as the tunica vaginalis, while the upper part seals off and loses continuity with the peritoneal cavity. In Daniel, the processus vaginalis remained continuous with the peritoneal cavity, giving rise to the hernia.
4. The inguinal hernia was of the 'indirect' variety because the sac passed through the deep inguinal ring.

CASE 36

1. The tumour had spread, via the lymphatics, to the regional lymph nodes in the mesentery and to the pre-aortic lymph nodes close to the superior mesenteric artery.
2. The tumour had spread via the lymphatic vessels rather than via the blood vessels. Tumour would have been expected in the liver if metastases had entered the tributaries of the portal vein.
3. From the pre-aortic lymph nodes, tumour probably passed via the intestinal lymph trunk and cisterna chyli to the thoracic duct.

Backflow from the thoracic duct to the hilar lymph nodes may account for the presence of tumour at this site. Although valves would have been expected to prevent backflow, Sapin and Borziak (1974) have demonstrated its occurrence.
4. At the lung root in the posterior mediastinum.
5. The left brachiocephalic vein and superior vena cava lie in the superior mediastinum. The superior vena cava passes through the pericardium to enter the right atrium.

CASE 37

1. The taeniae coli are three longitudinal bands of smooth muscle in the muscularis externa. They are a feature of the colon, but not of the appendix or rectum.
2. The diverticula tend to occur between the taeniae coli where the outer longitudinal layer of the muscularis externa is thinner and the bowel wall weaker.
3. The right and left subphrenic spaces are the regions of the peritoneal cavity inferior to the diaphragm, and are separated by the falciform ligament. They extend far posteriorly and tend to collect pus in a recumbent patient with peritonitis.
4. The spleen lies in the left hypochondriac region of the abdominal cavity. The stomach is separated from the spleen by the peritoneum of the greater sac, where the pus lay in the post-mortem subject.
5. The oxygen for the hepatocytes is largely supplied by the hepatic artery. The hepatocytes at the centre of the classical lobule are furthest from this source, and therefore most vulnerable to hypoxia.

CASE 38

1. Neurons from the appendix conveying the sense of pain pass mainly to the T10 spinal nerve. The sensation of pain at the umbilicus is a referred pain, the dermatome of T10 lying in this area.
2. As the inflammation advances, it involves the parietal peritoneum which is supplied by the nerves of the abdominal wall.
3. The somatic nerves are much more sensitive to pain than the visceral ones, so the pain became more severe as the parietal peritoneum became involved.
4. McBurney's point lies at the junction of the lateral and middle thirds of a line joining the right anterior superior iliac spine to the umbilicus.
5. Skin, superficial fascia, external oblique muscle, internal oblique muscle, transversus abdominis muscle, transversalis fascia, peritoneum.
6. The fibres of the muscular layers at McBurney's point run approximately in the following directions:
- External oblique muscle – downwards and medially.
- Internal oblique muscle – upwards and medially.
- Transversus abdominis muscle – horizontally.
7. McBurney's point is in the region where the fleshy fibres first become aponeurotic.
8. Ilio-inguinal nerve. Denervation of fibres of the internal oblique and transversus abdominis muscles, inserting via the conjoint tendon, could lead to laxity of the tendon and direct inguinal hernia.
9. The taeniae coli are followed to the base of the appendix.

CASE 39

1. The fundus of the gall bladder lies where the right side of the rectus abdominis muscle (linea semilunaris) meets the costal margin.
2. The bile duct runs on the posterior aspect of the head of the pancreas. The carcinoma had pressed on and occluded the bile duct. Jaundice occurred because bilirubin could not be excreted and was retained in the blood.
3. The bile leaves the liver by passing down the right or left hepatic duct, and then down the common hepatic duct. It then passes along the cystic duct to the gall bladder and enters the jejunum through the cholecystjejunostomy.

CASE 40

1. The male urethra has a prostatic part, where it lies in the prostate; a membranous part, where it traverses the urethral sphincter; and a spongiose part, where it passes through the corpus spongiosum of the penis.
2. The seminal colliculus is a rounded elevation on the posterior wall of the prostatic urethra. The prostatic utricle opens on its summit, and the ejaculatory duct has a slit-like opening to each side.
3. Each ejaculatory duct is formed by union of the ductus deferens with the duct of the seminal vesicle. It passes antero-inferiorly through the prostate to the prostatic urethra.
4. The prostate is composed of glandular tissue in a fibromuscular stroma.
5. The prostatic venous plexus lies between the capsule of the gland and the surrounding pelvic fascia. The bleeding, however, arises from vessels internal to the capsule.

CASE 41

The uterine tubes and uterus develop from the paramesonephric (Müllerian) ducts. The caudal ends of the ducts fuse to form the uterus. In Ms Waugh, the fusion was incomplete.

CASE 42

1. Urinary flow is controlled by the voluntary urethral sphincter around the membranous part of the urethra, assisted by pubo-vaginalis.
2. In sneezing, coughing and lifting, contraction of the abdominal muscles raises the intra-abdominal pressure, which cannot be resisted by the compromised sphincters.
3. The transverse cervical (cardinal) ligaments are condensations of fascia in the inferior part of the broad ligaments. They are principal supports of the uterus and tend to become lax in postmenopausal women, particularly those who have had children.
4. The lower part of the anterior vaginal wall is related to the urethra.
5. Loops of ileum in the recto-uterine pouch may cause a bulge (enterocele) at the posterior fornix. Similarly, the rectum may bulge into the lower vagina (rectocele)

CASE 43

1. The non-pregnant uterus is not palpable on routine abdominal examination.
2. The uterus was directed posteriorly with respect to the vagina.
3. The reason why pain occurs around the time of menstruation is not fully understood, but may be related to the stretching of structures by the menstrual process in the deposits of endometrial tissue.
4. No. If possible, vaginal examination would have been avoided, and the information sought by rectal examination instead, leaving the hymen intact.

CASE 44

1. The uterine tube lies in the upper border of the broad ligament.
2. Infundibulum, ampulla, isthmus and uterine part.
3. Fertilisation usually occurs in the ampulla.
4. The sigmoid colon and coils of ileum.
5. Through small abdominal incision(s), the uterine tubes are occluded by clips or ligatures and, depending on the technique, may be divided.
6. Probably, the blood in the peritoneal cavity irritated the diaphragm and pain was referred to the shoulder. The phrenic nerves, which are sensory to the diaphragm, and the lateral supraclavicular nerves, which supply the shoulder-tip, contain C4 fibres.

CASE 45

1. Human chorionic gonadotrophin (HCG). It is produced by the trophoblast.
2. Twelve weeks.
3. AFP occurs in high levels in fetal cerebrospinal fluid. In neural tube defects, the protein has ready access to the amniotic fluid. AFP crosses the placenta and can be detected in maternal serum. There are a number of explanations for elevated AFP levels in maternal serum. Further investigations, including ultrasound, are required to make the diagnosis.
4. The anterior and posterior neuropores close at 25 and 27 days, respectively, after conception.
5. Severe spina bifida.
6. The physiological hernia of midgut into the umbilical cord occurs from 6 to 12 weeks' gestation, approximately. The omphalocele in this fetus was abnormal.

CASE 46

1. Tibial collateral ligament.
2. Anterior cruciate ligament.
3. Rectus femoris, vastus lateralis, vastus intermedius and vastus medialis muscles. They are extensors of the knee.
4. On extending the knee, the anterior cruciate ligament becomes taut and acts as the fulcrum for the medial rotation of the femur, which locks the joint. The damage to the ligament has resulted in instability.

CASE 47

1. The medial meniscus is more commonly torn. It is less mobile than the lateral meniscus, being attached to the tibial collateral ligament. In addition, the action of the popliteus on the lateral meniscus reduces the risk of its entrapment between the lateral condyles of the femur and tibia.
2. The medial meniscus is a C-shaped incomplete ring of fibro-cartilage, wedge-shaped in cross-section. The horns of the meniscus attach to the intercondylar part of the tibial plateau, and peripherally it is attached to the tibial collateral ligament. The periphery of the meniscus is also attached to the margin of the head of the tibia, by part of the capsule known as the coronary ligament.
3. A piece of the torn meniscus was loose and able to be entrapped between the corresponding femoral and tibial condyles.
4. The patella is a sesamoid bone within the quadriceps tendon. The quadriceps muscles attach to the tibial tuberosity by the patellar ligament.
5. The medial compartment is the portion of the joint between the medial condyles of the femur and tibia. The space seen on radiology of a healthy limb represents the radiolucent articular cartilage of the femoral and tibial condyles and the intervening medial meniscus.

CASE 48

1. At the groin, the femoral vessels lie in the femoral triangle. The femoral vein enters the triangle from the subsartorial (adductor) canal and at first lies behind the femoral artery. The upper 3–4 cm of the vein lie medial to the artery in the femoral sheath. The femoral vein is continuous with the external iliac vein behind the inguinal ligament.
   The femoral artery passes behind the inguinal ligament at the mid-inguinal point, midway between the anterior superior iliac spine and the pubic symphysis; the vein is immediately on its medial side.
2. Intravenous injections are usually given into the superficial veins of the upper limb. In the lower limb, there are many superficial veins which may be visible through the skin. The larger veins are the long (great) and short (small) saphenous veins. The long saphenous vein is particularly accessible anterior to the medial malleolus.
3. Arrangement of deep veins of lower limb:
   • Below the knee, arteries are accompanied by venae comitantes.
   • The posterior tibial veins receive blood from the calf muscles, especially from the venous plexus in the soleus muscle, as well as from the peroneal veins.
   • The anterior tibial veins drain the front of the leg and pass through the opening in the upper part of the interosseous membrane between the tibia and fibula.
   • The anterior and posterior tibial veins unite and form the popliteal vein.
   • The popliteal vein leaves the popliteal fossa through the hiatus in the adductor magnus muscle and is continuous with the femoral vein in the subsartorial canal.
   • In the lower part of the femoral triangle, the femoral vein is joined by the profunda femoris vein, which drains the thigh.
   • The femoral and popliteal veins receive the long and short saphenous veins, respectively.

• Perforating veins carry blood from the superficial to the deep veins.
4. Factors facilitating return of blood:
   • Valves prevent blood from flowing back into the deep veins and escaping into the superficial veins.
   • Contraction of the calf muscles, particularly the soleus muscle, pumps blood from venous plexuses within the muscles.
   • Contraction of muscles and arterial pulsations compress veins and milk the blood along.
   • During inspiration, the fall in intrathoracic pressure and the rise in intra-abdominal pressure aids venous return to the heart.

CASE 49

1. Gastrocnemius and soleus muscles.
2. The soleus muscle forms a tripartite muscle mass with the lateral and medial heads of the gastrocnemius muscle.
3. The bursa which separates the tendo calcaneus from the upper part of the posterior surface of the calcaneus.
4. In this part of the gait cycle, flexion of the ankle is most marked and rapid.
5. When standing, the body's centre of gravity falls in front of the ankle joint. The soleus muscle acts to prevent the body falling forwards. Thus, there is tension in the tendo calcaneus, but little movement.

CASE 50

1. Lymphangitis – inflammation of the lymphatic vessels.
2. The superficial and deep inguinal lymph nodes lie at the groin. The superficial nodes consist of an upper and a lower group. The upper group is 5–6 nodes, just below and parallel to the inguinal ligament. The lower group is 4–5 nodes along the terminal part of the great saphenous vein. The deep nodes are 1–3 nodes alongside the femoral vein.
3. The popliteal lymph nodes receive lymph from the back and lateral side of the calf, including the lateral side of the heel.
4. The skin of the lateral toes drains to the lower group of super-ficial inguinal lymph nodes.
5. Other sites drained by lymph nodes:
   • The skin of the lower limb drains to the lower group of super-ficial inguinal lymph nodes.
   • The upper group receives lymph from the anterior abdominal wall and perineum.
   • The deep inguinal lymph nodes drain deep tissues of the limb, and also receive lymph from the popliteal lymph nodes and some of the superficial inguinal nodes, as well as from the glans penis or glans clitoridis.

CASE 51

1. The popliteal, dorsalis pedis and posterior tibial pulses are readily palpable in healthy individuals.
2. Identification of features A to D:
   • A = (Superficial) femoral artery.
   • B = Popliteal artery.
   • C = Anterior tibial artery.
   • D = Posterior tibial artery.

3. Some branches of the profunda femoris artery are seen on both sides – on the left more than the right. On the left side, a little contrast medium is seen in the (superficial) femoral and popliteal arteries; this has probably arrived there via anastomoses with the profunda femoris artery.

CASE 52

1. A bursa is a flattened sac of synovial membrane which reduces the friction of moving muscles or tendons. The two layers of synovial membrane are apposed and separated by a film of synovial fluid.
2. The trochanteric bursa lies between the gluteus maximus muscle and the greater trochanter.
3. The gluteus maximus muscle is an extensor of the hip joint.
4. In walking, the hamstrings are responsible for extension of the hip. The gluteus maximus muscle is used when more powerful extension is required, such as climbing stairs.
5. If the subject stands on one leg, the unsupported side of the pelvis rises because of the contraction of the contralateral gluteus medius and minimus muscles. In walking, this allows the un-supported limb to swing clear of the ground.
   With paralysis of the contralateral hip abductors, hip dislocation, or (as in Mr Goldie's case) a painful hip, this action does not occur, and the unsupported side of the pelvis sags. This is a positive Trendelenburg's sign.

CASE 53

Prepatellar bursitis (housemaid's knee); inflammation of the prepatellar bursa.

CASE 54

1. The sensory and motor speech areas were not involved. Mr McRoberts could understand language and name objects. He had dysarthria rather than dysphasia; the problem was one of articulation as a result of the motor deficit. In addition, the sensory language and motor speech areas are almost always found in the left cerebral hemisphere.
2. The motor neurons supplying the frontalis and orbicularis oculi muscles receive neuronal input from both motor cortices. The motor neurons supplying muscles in the lower face receive projections from the contralateral cortex only.
3. Almost all the musculature of the left side was affected. Muscle function in the upper face was spared. Muscle tone was in-creased. Reflexes were brisk. Plantar reflexes were upgoing.
4. The stroke was the result of the infarction in the posterior limb of the internal capsule. The principal motor and sensory pathways pass through this narrow region of white matter; damage at this site has disastrous results.

CASE 55

1. The right abducens (sixth cranial) nerve.
2. The cerebral aqueduct lies in the midbrain.

3. Choroid plexus produces CSF and is found in the lateral, third and fourth ventricles.
4. CSF percolates through the subarachnoid space between the arachnoid and pia mater.
5. Arachnoid villi invaginate the cranial venous sinuses, especially the superior sagittal sinus, and enable resorption of CSF into the bloodstream.
6. CSF passes through the midline and lateral foramina in the roof of the fourth ventricle.
7. The lateral and third ventricles were likely to have been disten-ded.

CASE 56

1. The subdural space is the potential space between the dura and arachnoid maters; it usually contains only a thin film of tissue fluid.
2. The location of the haematoma and the slow onset of symptoms indicate that one or more of the cerebral veins, which cross the subdural space, must have torn. This probably occurred when Mrs Gibson hit her head on the table.
3. Papilloedema indicates that the intracranial pressure is raised. The increase in pressure forces CSF along the extension of the subarachnoid space, between the optic nerve and its sheath, resulting in swelling of the disc.
4. The features on the CT scan are:
 • A = Anterior part of falx cerebri.
 • B = Anterior horn of left lateral ventricle.
 • C = Head of caudate nucleus.
 • D = Septum pellucidum.
 • E = Posterior horn of left lateral ventricle.
 • F = Posterior part of falx cerebri.

CASE 57

1. The thalamus is composed of grey matter and is deeply placed in the forebrain adjacent to the third ventricle.
2. All modalities of sensation, including vision and hearing, are integrated in the thalamus.
3. The ventral posterior nucleus.
4. The primary somaesthetic cortex – areas 1, 2 and 3 of Brodmann.
5. The blood supply to the thalamus mainly comes from fine branches of the posterior and posterior communicating arteries. These penetrate the posterior perforated substance and pass superiorly to the thalamus.

CASE 58

1. The middle cerebral artery arises from the circle of Willis and enters the lateral fissure where it divides into its upper and lower branches, which supply various regions, as follows:
 • The upper branch supplies the postero-inferior part of the frontal lobe and the antero-inferior part of the parietal lobe.
 • The lower branch supplies the postero-inferior part of the parietal lobe and the superior and anterior parts of the temporal lobe.

- Central branches arise from the proximal part of the middle cerebral artery to supply the corpus striatum, internal capsule and thalamus.
2. The damage is consistent with occlusion of the middle cerebral artery distal to its central branches.
3. The internal capsule seemed intact, but the area of motor cortex devoted to the left upper limb had been lost. The lower limb is served by the uppermost parts of the sensorimotor strip. This lies in the parasagittal cortex, supplied by the anterior cerebral artery, and had been spared.

   The head is represented in the lowest parts of the motor cortex which also had been lost. There was no history of facial palsy and, given that the patient was a child when he had contracted meningitis, it was assumed that other areas of the cortex had provided control of the facial muscles.
4. The degenerative changes in the corpus striatum and thalamus probably resulted from the loss of cortical connections.
5. The likely cause of the infarction was septic thrombosis of the artery resulting from meningitis.
6. In a healthy brainstem and spinal cord, corticospinal fibres from the right motor cortex lie in sections at the following levels:
- Midbrain – in the middle part of the right basis pedunculi.
- Pons – among the pontine nuclei and transverse pontine fibres of the right side of the basilar part of the pons.
- Upper medulla – in the right medullary pyramid.
- Cervical spinal cord – in the left lateral white funiculus (column), and in the right ventral white funiculus (column).
7. The pyramidal decussation lies in the lower medulla.

CASE 59

1. The disc consists of the anulus fibrosus, composed of concentric layers of fibrocartilage, which surrounds a gelatinous centre, the nucleus pulposus. In a prolapsed disc, the nucleus pulposus herniates through the anulus fibrosus. This usually occurs at the posterolateral aspect of the anulus – its weakest point.
2. S1/S2 on the right side.
3. L5 dermatome.
4. L5 nerve.
5. The ventral and dorsal rami emerge from the anterior and posterior sacral foramina, respectively.
6. At the intervertebral foramina; the ventral and dorsal roots are surrounded by a common sleeve of dura as they cross the epidural space.
7. The roots of L5 and S1 lie close to the intervertebral disc between L5 and S1 vertebrae. Herniation of the disc tends to press on S1 which is more medially placed.

CASE 60

1. The intervertebral disc is a symphysis. The zygapophyseal (facet) joints are synovial. In the cervical region, the bodies have curved lateral edges which meet at small synovial joints.
2. The normal curvatures are a cervical lordosis, thoracic kyphosis, lumbar lordosis, and sacro-coccygeal kyphosis. The sacro-coccygeal kyphosis reflects the shape of the bones; the other curvatures are mainly due to the shape of the intervertebral discs.
3. Between typical cervical vertebrae, and at the atlanto-occipital joint, the movements are flexion, extension and lateral flexion. Rotation occurs primarily at the atlanto-axial joint.
4. On each side, the vertebral artery is transmitted through the foramina transversaria of the upper six cervical vertebrae. It then enters the foramen magnum by passing posteriorly round the lateral mass of the atlas.
5. Assessment of the sufficiency of the vertebral artery is carried out with great care, and is not done if there is any possibility of instability of the cervical spine. The patient lies supine on a couch with the head over the end; the physiotherapist provides support. The neck is then extended and laterally flexed; dizziness suggests disease of the contralateral (opposite side) artery which is stretched by the procedure. If dizziness (vertigo) occurs, the vestibular apparatus should be tested to exclude inner ear disease before a firm diagnosis is made.
6. It was unclear why the headache was felt in the occipital region. The pain may have been related to the pressure on the dorsal ramus of C2, whose main branch, the greater occipital nerve, supplies the skin in the region.

CASE 61

1. Zygapophyseal joints are found between adjacent vertebrae throughout the cervical, thoracic and lumbar regions of the vertebral column.
2. Zygapophyseal joints occur between the inferior and superior articular processes of adjacent vertebrae.
3. Synovial joints.
4. The movements in the lumbar spine are flexion, extension and lateral flexion; rotation is minimal.
5. Involvement of L4 was suggested by the position of the back pain, and from the history of pain referral to the lower thigh, front of leg, and foot towards the big toe.

   The distribution of the referred pain corresponds in a general way to the dermatome of L4. Much of the innervation of an L4/L5 zygapophyseal joint comes from L4, but it also receives branches from L3 and possibly other nerves. Pain from the joint may therefore have been referred to more than one dermatome.

INDEX

Numbers are question and answer numbers.